D0152253

Second Edition

Elementary Number Theory

Charles Vanden Eynden

Illinois State University

Long Grove, Illinois

For information about this book, contact:
Waveland Press, Inc.
4180 IL Route 83, Suite 101
Long Grove, IL 60047-9580
(847) 634-0081
info@waveland.com
www.waveland.com

Reissued 2006 by Waveland Press, Inc.

ISBN 1-57766-445-0

Printed in the United States of America

7 6 5 4 3 2 1

Credits

Page 11: Photo of Euclid. Courtesy of the Archives Center, National Museum of American History. Negative No. 57,973.

Page 46: Photo of Fermat. North Wind Picture Archives.

Page 68: *Machine Performs Difficult Mathematical Calculations* Carnegie Institution of Washington News Service Bulletin vol. III, no. 3, 1933. Reprinted with permission of the Carnegie Institution of Washington.

Page 75: Photo of Euler. Courtesy of the Archives Center, National Museum of American History. Negative No. 13,189-A.

Page 95: *Antiche Danze ed Arie per liuto*, Respighi. © Casa Ricordi, reproduced by permission.

Page 96: *Mersenne and Fermat Numbers*, Raphael M. Robinson. Reprinted from *Proceedings of the American Mathematical Society*, volume 5 (1954).

Page 97: *Prime Number Record Broken*. Reprinted from Agence France-Presse, November 17, 1978.

Page 115: Photo of Gauss. Courtesy of the Archives Center, National Museum of American History. Negative No. 46,834-N.

Page 155: *Opening the Trapdoor Knapsack*, Time Magazine, October 25, 1982. ©1982 Time Inc. Reprinted by permission.

Page 156: Republished with permission of Boston Globe, *NM Scientist Cracks Code*, Edward Dolnick, November 6, 1984; permission conveyed through Copyright Clearance Center, Inc.

Page 158: Photo of Lagrange. Courtesy of the Archives Center, National Museum of American History. Negative No. 46,834-N.

Page 194: Photo of Hardy. Reprinted with permission of Cambridge University Press.

Page 194: Photo of Ramanujan. Reprinted with the permission of Cambridge University Press.

Page 233: Photo of Erdos. Courtesy of The MacTutor History of Mathematics Archive, University of St. Andrews, Scotland.

Contents

Preface

This book was written with four goals in mind:

1. To be easy for the student to read and easy for the professor to teach from.

2. To be suitable for a wide variety of audiences.

3. To involve the student actively in the construction of proofs.

4. To develop in the student a self-confidence with respect to mathematics that will foster interest and success.

To achieve these goals I have set off from the text definitions and examples, as well as theorems, so that they are easy to find and so that the central topics of each section are apparent. In addition, each section is followed by a set of problems divided into three parts. The problems in part A are generally straightforward calculations using techniques covered in the section. Problems in part B include calculations needing some ingenuity or new ideas and standard proofs. Part C problems call for harder proofs. (Of course, the classification of the problems, especially those in parts B and C, calls for a judgment with which the reader may sometimes disagree.) Also, to give a taste of how advances in mathematics really are made, a large number of true-false problems are given. These foster experimentation and a healthy skepticism.

I believe number theory is an ideal subject for introducing the craft of proof construction, by reason of its concreteness and familiarity. However, proofs cause great anxiety in many students; there is nothing safe about finding a proof, the way doing a calculus problem is safe (one merely learns the technique, then applies it). Anxiety with respect to proofs makes success less likely in most cases. Therefore, the number of computational problems in this book exceeds that of most number theory texts. A substantial portion of problems of a computational or algorithmic nature provide the student with a floor of activities he or she can feel confident about. Computational problems also lead to understanding of the theory behind them. Moreover, the large supply of easier problems also enables the professor to make an assignment every day, even when not much is covered in class.

To further reduce anxiety about proofs, the ideas behind a proof are explained whenever possible. The student is shown that theorems and proofs do not spring from the pens of mathematicians in a perfect form, ready to be printed in books, but rather are the result of trial and error and evolutionary improvement, like most high-level human accomplishments. Fortunately, number theory has a rich history, and many historical details are included, both for their intrinsic interest and to show that mathematics is an activity of people.

Historically in mathematics, a problem arises, and then (sometimes) it is solved. In texbooks, however, it is usual to give the solution, followed by applications. This order of presentation is compact and logical, but the historical sequence is generally better pedagogy. Whenever possible, I have tried to set up theorems as solutions to naturally arising problems or conjectures. Occasionally this approach requires introducing a concept before it can be developed fully; an example is the function $\tau(n)$ in Section 2.1.

I have aimed for a writing style that is informal yet to the point. If a theorem needs a few words to put it in context or an example to illustrate either its proof or application, it is provided; but I have tried to avoid saying what is obvious. Section heading and other titles have been chosen to be informative rather than clever. My aim has been to talk straight to those who will read the text and provide a convenient organization for the rest.

Several topics receive special emphasis in this book. A very short introduction to logic and mathematical proof is given in Chapter 0. A section on mathematical induction is provided in Chapter 1 in order to get across the idea that induction is more than an algorithm for proving summation formulas. In addition, the use of the sigma and pi notations for sums and products is treated at length, especially with respect to the way these notations are used in number theory. Chapter 6 opens with a section on countability as it applies to the rational and real numbers. Although this is somewhat unusual in a number theory text, I find this material puts continued fractions in a richer context, and undergraduates are sometimes unfamiliar with it. Later sections do not depend on this one. An increasingly important application of number theory is the RSA method of public key cryptography, and it is presented in a way that enables students to actually use it with numbers that can be handled with a scientific calculator. Because of student questions when teaching the method for a number of years, I have added to this second edition a section on factorization and primality testing.

Theorems have been numbered consecutively in chapters, so that Theorem 1.3 is the third numbered theorem in Chapter 1. Although such numbering is sometimes convenient because of later references, students can hardly be expected to remember these numbers. Thus whenever possible I have used names for theorems (for example, the Euclidean algorithm). Results used just to illustrate proof techniques or followed by more general versions are called *Propositions.*

Depending on the speed with which topics are covered and what is expected from the student, it is possible to use this book for a wide range of courses, from a number-theory-appreciation treatment emphasizing history and type A problems to a proof-oriented course preparing the student for graduate courses. A minimal course would probably include the first four chapters, but even so, Sections 3.6, 4.5, and 4.6 and some topics (such as Fibonacci numbers, dividing exactly, and amicable, abundant and deficient numbers), could easily be skipped. The last three chapters are largely independent of each other, except that in Chapter 7 the treatment of sums of two and four squares depends on Section 5.3 and the presentation of Pell's equation depends on Section 6.3.

Acknowledgments

This book has been improved by the corrections and suggestions of many conscientious readers. Over the years my number theory colleagues at Illinois State University, Larry Eggan, Earl Ecklund, Andy Pollington, and Roger Eggleton, have provided much help and encouragement.

The following reviewers can take credit for much of what is good in the book and blame for none of my stubborn idiosyncrasies:

First Edition

Gerald L. Alexanderson, University of Santa Clara
Bruce C. Berndt, University of Illinois
Roger Entringer, University of New Mexico
Joel K. Haack, Oklahoma State University
Neal Hart, Sam Houston State University
Padmini T. Joshi, Ball State University
Carole B. Lacampagne, University of Michigan–Flint
Arne Magnus, Colorado State University
Paul J. McCarthy, University of Kansas
Charles J. Parry, Virginia Polytechnic Institute and State University
William H. Richardson, Wichita State University
Daniel H. Saracino, Colgate University
Lawrence C. Washington, University of Maryland
William A Webb, Washington State University

Second Edition

Todd Cochrane, Kansas State University
Joe Dennin, Fairfield University
David Gove, California State University Bakersfield
Stan Gudder, University of Denver
John Johnson, George Fox University
Mike Miller, Western Baptist College
Tom Nelson, Sonoma State University

These lists attest to the care given this book by John R. Martindale, Judith Kromm, Maggie Rogers, and Jill Peter in addition to many creative ideas for its improvement.

Finally, it has been my pleasure to benefit from the lectures and writings of many fine number theorists, but especially those of Ivan Niven.

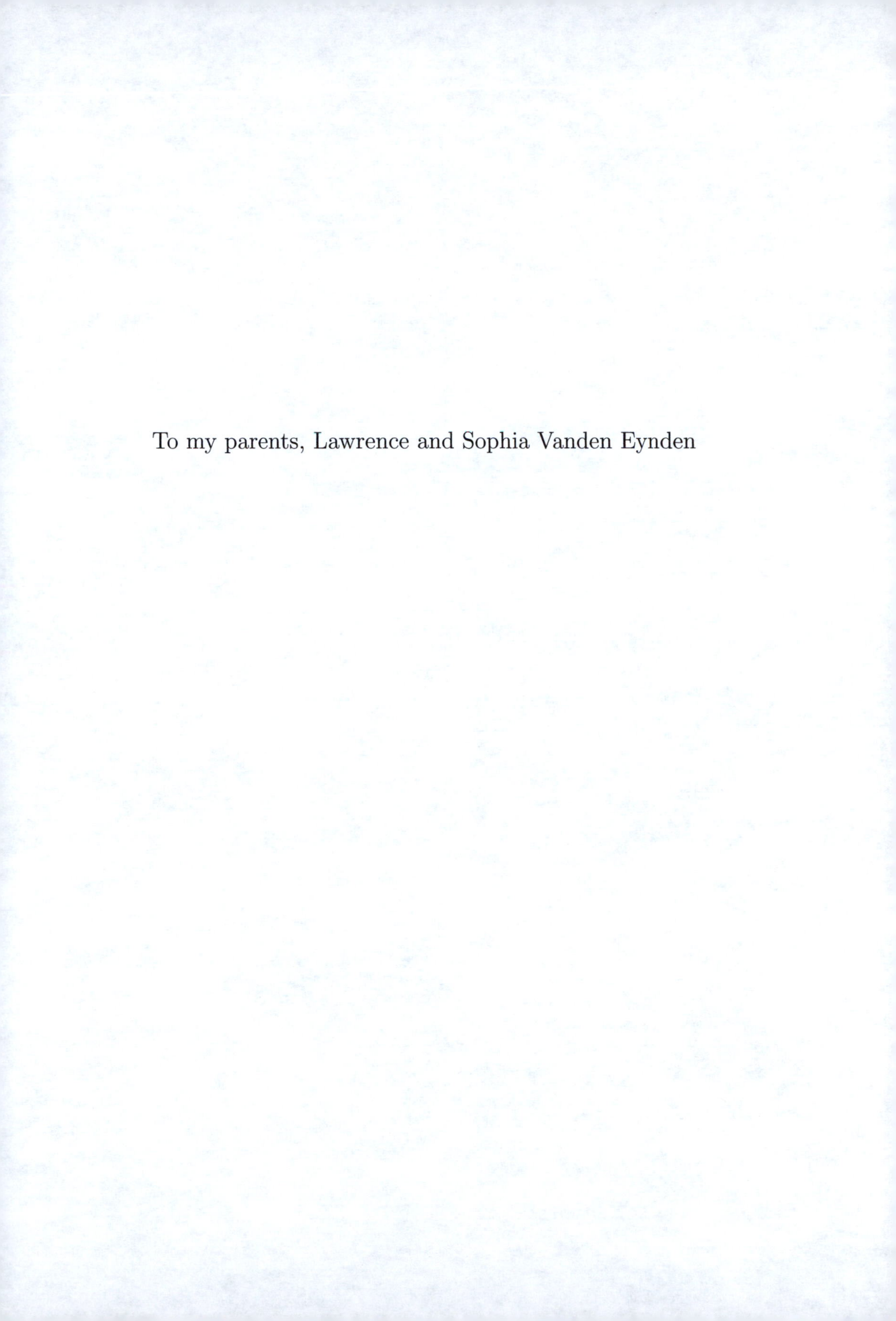

To my parents, Lawrence and Sophia Vanden Eynden

Chapter 0

What Is Number Theory?

No completely satisfactory definition has ever been given, but roughly, number theory is the mathematical treatment of questions related to the integers, that is, the numbers $0, 1, -1, 2, -2, \ldots$. Investigating such questions has often led mathematicians into other fields, however, such as complex numbers, geometry, and abstract algebra.

Number theory may very well be the best subject for a student trying to learn what constitutes a mathematical proof, and to construct proofs. It combines the advantages of concreteness and familiarity. The integers possess a concreteness, a realness, that more abstract concepts (vector spaces, for example) do not. This concreteness enables us to form hypotheses about the integers and test them by actual experiment.

Of course, the integers are familiar to us from our earliest introduction to arithmetic. We have manipulated them thousands of times, and have formed an intuitive sense of many of their laws. This intuition carries some danger with it, because it may be hard to see the necessity for a proof to verify patterns that we have confirmed by instances over and over again since grade school. Although intuition can be of immense help in mathematics, no accumulation of special cases is sufficient to prove a general proposition.

This book contains a large quantity of exercises, mainly of two types: computations and proofs. There are many more computational problems than in most number theory texts. The student who finds a particular section hard to navigate is advised to self-administer large doses of these computational problems; they lead to understanding.

Proofs are another matter. A calculus student may feel he or she should be able to do any assigned problem by applying the standard methods of the subject. On any particular day, however, a person may or may not be able to construct a proof of a given mathematical fact. There are no infallible methods. If there were, the proof would be unnecessary; one could merely say "Apply method A." Thus no one should expect to plow through a set of proof problems like a list of differentiation problems, knocking them off one after the other. A success rate of 100% is too much to expect. Sometimes the right idea comes at once, sometimes after an hour's work, sometimes after a night's sleep, sometimes never.

Most assertions to be proved are **implications**, that is statements of the form "if P, then Q," where P and Q are themselves statements. The statement P is called the **hypothesis** and Q is called the **conclusion**. An example is:

S: If Saturday is your birthday, then I bought you a present.

Here P is the statement "Saturday is your birthday," and Q is the statement "I bought you a present."

The implication "if P, then Q" is considered to be true unless P is true but Q is false. Suppose I make the statement to one of my daughters "If Saturday is your birthday, then I bought you a present." My statement is truthful in each of the following cases:

- Saturday is her birthday and I bought her a present.
- Saturday is not her birthday and I didn't buy her a present.
- Saturday is not her birthday but I bought her a present anyway.

I have lied only if Saturday is her birthday but I didn't buy her a present. The meaning of "if P, then Q" is summarized in the following truth table.

P	Q	if P, then Q
true	true	true
true	false	false
false	true	true
false	false	true

Notice that "if P, then Q" is not the same as "if Q, then P"; and one of these implications may be true and the other false. In our example, the second statement would be the assertion:

S': If I bought you a present, then Saturday is your birthday.

The statement "if Q, then P" is called the **converse** of the statement "if P, then Q."

Actually in mathematics almost all statements are more complicated than those of the above example because they involve one or more variables. These statements might have the form "for all x, $P(x)$," or "for some x, $P(x)$, where $P(x)$ is some statement involving the variable x, whose truth value depends on x. An example (talking about real numbers) is "if $x > 4$, then $x > 2$." What is really meant here is

T: For all real numbers x, if $x > 4$, then $x > 2$.

Although *converse* is defined in most mathematics books as it was above, its actual usage is generally different. For example, if a mathematician talks about the converse of the statement T, he or she probably means

T': For all real numbers x, if $x > 2$, then $x > 4$.

In this case T is true, but T' is false, as the **counterexample** $x = 3$ shows.

There are many sets of True-False exercises in this book. The idea is to prove those statements that are true and disprove the others. The disproofs ordinarily take the form of a counterexample. The more concrete and definite the counterexample is, the better. For example, consider the statement

If $a \geq b$, then $ac \geq bc$.

This could be disposed of by

FALSE. Let $a = 2$, $b = 1$, and $c = -1$. Then $a = 2 \geq 1 = b$, but $ac = -2$ and $bc = -1$. Thus $ac \geq bc$ is false.

This is superior to

FALSE. If $a \geq b$ and c is negative, then $ac < bc$.

The second disproof is inferior in two respects. First, it is more general than it has to be, invoking a theorem about inequalities when all that is really needed is that $-2 < -1$. Second, it is itself false. What if $a = b = 0$ and $c = -1$? Then $a \geq b$ and c is negative, yet $ac < bc$ is false. True enough, the second disproof is valid except in the special case that $a = b$. But that is enough to invalidate it. Mathematics is a very picky subject.

It should be mentioned that a purely numerical counterexample does not always suffice to disprove a statement. Consider

There exists x such that $x > y$ for all y.

It is not good enough to say

FALSE. Let $x = 100$. Then if $y = 200$, $x > y$ is false.

After all, nobody said x was 100. It must be shown that *no* x satisfies the statement.

FALSE. Let any x be given. Set $y = x + 1$. Then $x > y$ is false.

The above is a satisfactory disproof.

Sometimes when you are trying to prove something and are stumped, it helps to look for a counterexample. If you can see why you can't find one, you are on your way to a proof.

It is a ground rule that in proving an exercise any previous exercise or result of the text may be assumed unless the contrary is expressly stated.

Some problems and many theorems contain the words "if and only if." Proving that P if and only if Q (where P and Q are statements) usually breaks into two parts:

- P if Q (i.e., whenever Q is true, so is P), and

- P only if Q (i.e., whenever P is true, so is Q).

As an example, consider the familiar statement:

> If A, B, and C are real numbers with $B^2 - 4AC > 0$ and $A \neq 0$, then $Ax^2 + Bx + C = 0$ if and only if x is

$$\frac{-B + \sqrt{B^2 - 4AC}}{2A} \quad \text{or} \quad \frac{-B - \sqrt{B^2 - 4AC}}{2A}.$$

The "if" part of the statement says that each of the two numbers satisfies the quadratic equation, and could be proved by direct substitution. The "only if" part says these are the only solutions, and is usually proved by completing the square on the original equation.

Sometimes we may say that a proof is "indirect," or a "proof by contradiction." This simply means that the scheme of the proof is to assume that the statement to be proved is false. Then contradictory statements are proved. From this we conclude that the statement to be pròved must be true. As an example we offer (with a few comments enclosed in brackets) the standard proof that $\sqrt{2}$ is irrational (that is, not the quotient of two integers).

Proposition *The number $\sqrt{2}$ is irrational.* [The words "the number" were inserted just to avoid starting a sentence with a symbol.]

Proof. We will give an indirect proof. [It is always wise to let the reader in on the type of proof we have in mind.] Suppose $\sqrt{2}$ is rational. [Here is where we assume the statement to be proved is false.] Suppose

$$\sqrt{2} = \frac{a}{b},$$

where a and b are integers and the fraction a/b is in lowest terms. [When the new symbols a and b are introduced we are careful to explain what they represent.]

Squaring both sides leads to

$$2b^2 = a^2.$$

From this equation we see that a^2 is even. Thus a must be even, for if $a = 2k+1$, then

$$\begin{aligned} a^2 &= (2k+1)^2 = 4k^2 + 4k + 1 \\ &= 2(2k^2 + 2k) + 1, \end{aligned}$$

which is clearly odd. [This part of the proof illustrates one of the hardest questions involved in presenting an argument, namely, how much detail to include. If this proof were given later in the course, it would probably be reasonable to assume the reader would know that the square of an odd integer had to be odd. But since we are only in Chapter 0 we have included the explanation beginning with "for if $a = 2k + 1$." It is always better to err in the direction of giving too much detail.]

Since a is even, let $a = 2c$, where c is an integer. [The new symbol c is explained.] Then $2b^2 = 4c^2$, and so

$$b^2 = 2c^2.$$

[Have we been too optimistic in assuming the reader will see we substituted $a = 2c$ into the equation $2b^2 = a^2$?]

Thus we see that b^2 is even, and so b must be even, as above. [No sense in repeating the argument we just made for a.]

But this contradicts the assumption that a/b was in lowest terms, since we have seen that both a and b are even. This contradiction proves that $\sqrt{2}$ is irrational. □

Note that in the above proof we had to indicate specifically that the variables a, b, and c were integers. We will save time in the future by adopting the following convention:

CONVENTION Since the integers will be our main concern, we will make the convention that all lower case italic letters (a, b, c, ... , x, y, z) will denote integers.

Number theory is quite an old subject, and some of the theorems in this book may be found in the writings of Euclid, who lived about 300 years before Christ. It might be thought that all the natural questions that arise about the integers have been solved for a long time, but this is not the case. The field abounds in problems that can be stated simply but that no one has been able to solve for centuries. We will list a few of these here; later in the book historical details of some of them will be given.

A few simple definitions are needed. These will be given quickly here and repeated later when the theory is developed in a more systematic way. We say that the integer a is a **divisor** of the integer b if b/a is an integer; for example, 3 is a divisor of 15, but 3 is not a divisor of 16. We say that the integer p is **prime** if $p > 1$ and the only positive divisors of p are 1 and p itself; the first three primes are 2, 3, and 5.

Unanswered Questions

1. Many examples of pairs of primes differing by 2 can be found, such as 3 and 5, 5 and 7, 11 and 13, and 17 and 19. Are there infinitely many of these? [the twin prime conjecture]

2. It seems to be possible to write each even number greater than 2 as the sum of two primes. For example, $4 = 2 + 2$, $6 = 3 + 3$, $8 = 3 + 5$, and $10 = 3 + 7$. Can this always be done? [the Goldbach conjecture]

3. Are there infinitely many primes of the form $n^2 + 1$, such as $1^2 + 1 = 2$, $2^2 + 1 = 5$, $4^2 + 1 = 17$, and $6^2 + 1 = 37$?

4. Are there infinitely many primes of the form $2^n - 1$, such as $2^2 - 1 = 3$, $2^3 - 1 = 7$, and $2^5 - 1 = 31$? [Mersenne primes]

5. Are there infinitely many primes of the form $2^n + 1$, such as $2^1 + 1 = 3$, $2^2 + 1 = 5$, and $2^4 + 1 = 17$? [Fermat primes]

6. An integer is said to be **perfect** if it is the sum of its positive divisors other than itself. An example is $6 = 1 + 2 + 3$. Are there infinitely many perfect numbers?

7. Are there infinitely many odd perfect numbers? Is there even one?

8. The decimal expansion of $1/7$ is $0.142857142857\ldots$, where the six digits 142857 repeat forever. Are there infinitely many integers n such that the smallest set of repeating digits in the decimal expansion of $1/n$ has length $n - 1$?

Lest the reader be left with too pessimistic a view of number theory, we will list several results that *can* be proved. Most of these will be found in this book, but a few have proofs too difficult to include.

9. The equation $x^2+y^2 = z^2$ has infinitely many solutions in positive integers with no common divisor greater than 1. In fact, in some sense the solutions to this equation have been completely characterized; see Section 7.1. [Pythagorean triples]

10. There are infinitely many primes. [Euclid]

11. In fact, the series $1/2+1/3+1/5+\cdots$ with the reciprocals of the primes as terms diverges. [Euler]

12. On the other hand, if we restrict the series in the previous item to primes differing by 2 from another prime (see question (1)), the series $1/3+1/5+1/7+1/11+1/13+\cdots$ converges. (Of course, this may be a finite sum.) [Brun's theorem]

13. If a and b are positive integers with no common divisor greater than 1, then the arithmetic progression a, $a + b$, $a + 2b$, $\ldots$ contains infinitely many primes. An example, with $a = 3$ and $b = 10$, is that the sequence 3, 13, 23, 33, 43, $\ldots$ contains infinitely many primes. [Dirichlet's theorem]

14. If $2^n - 1$ is prime, then n itself must be prime. (See question (4).)

15. If $2^n + 1$ is prime, $n > 0$, then n must be a power of 2. (See question (5).)

16. If n is an *even* perfect number, then n has the form $2^{k-1}(2^k - 1)$, where $2^k - 1$ is prime. Conversely, if $2^k - 1$ is prime, then $2^{k-1}(2^k - 1)$ is perfect. (See question (6).) [Euler, Euclid]

17. If the smallest set of repeating digits of $1/n$ has length $n-1$, then n is prime. (See question (8).)

18. Each positive integer n can be written in the form $w^2+x^2+y^2+z^2$, where w, x, y, and z are nonnegative integers. [Lagrange]

Fermat's Last Theorem

We have saved the biggest success for last, the recent proof by the British mathematician Andrew Wiles of what is known as Fermat's last theorem. It concerns an equation related to that defining Pythagorean triples (see statement (9)), namely,

$$x^n + y^n = z^n.$$

Here x, y, and z are supposed to be positive integers and n an integer *greater* than 2. Nobody has ever found a solution to this equation, making an odd contrast to the case when $n = 2$, for which there are infinitely many solutions.

About 1637 the Frenchman Pierre de Fermat, a lawyer and probably the last great amateur mathematician, wrote in a marginal note in his copy of the *Arithmetica*, a book on number theory by the Greek mathematician Diophantus, that he had found a wonderful proof that the equation had no solutions, but that there was not room to write it down. The statement was found and published by Fermat's son after his death, and came to be known as **Fermat's last theorem**, even though a "theorem" really should have a proof.

Mathematicians proved cases of the theorem for particular values of n, and by the early 1990s all values up to more than 100,000 had been accounted for. Nevertheless no general proof was found. Even though the problem itself was of no known practical importance, the research it sparked opened many new and significant areas of mathematics.

Finally in 1993 Wiles announced a proof. He had read about the problem in a library book at the age of 10, and worked on it ever since. The bug for proving the theorem caused him to become a professional mathematician, and he spent 9 years on the problem in his attic before announcing a proof.

Some skepticism was in order, because proofs of Fermat's last theorem had been claimed before. As recently as 1988 a Japanese mathematician, Yoichi Miyoaka, announced a possible proof of the theorem, which was reported in major newspapers and magazines. Within weeks doubts arose, however, and in the end the whole thing turned out to be a false alarm.

The problem entered a new stage on the 21st, 22nd, and 23rd of June, 1993 when Andrew Wiles gave a series of three lectures at the Newton Institute in Cambridge, England. During the last of these he announced a proof of Fermat's last theorem. This announcement caused quite a stir, even among the general public. Wiles, often seen in khakis and tennis shoes, turned down $750 to appear in a Gap ad. He showed his "proof" to only a small group of number theorists, and skepticism among the wider mathematical world began to grow. Finally in December, Wiles announced that while he had fixed a number of difficulties arising in writing up his arguments in detail, one remained that he could not resolve. He essentially took back his claim of a proof.

In the months that followed, Wiles and a few others familiar with his methods worked feverishly to plug the hole. Finally at the end of the summer of 1994, Wiles apparently found a way around the problem, simplifying his earlier arguments in the process. At present it seems generally accepted that Fermat's last theorem has been proved, even though the proof is so long and deep that only a small fraction of mathematicians have the background to check it personally.

The Goldbach Conjecture

Even though most of the questions in the first list we gave have been unanswered for many years, even centuries, progress continues to be made on some of them. We will consider the history of problem (2), which asks whether each even integer greater than 2 can be written as the sum of two primes. On the basis of numerical evidence, the German-born Christian Goldbach conjectured that the answer was yes in a letter to the Swiss mathematician Euler in 1742; and this guess is now known as the **Goldbach conjecture**. It was first published by the English mathematician Edward Waring in 1770, but little progress was made on proving the conjecture until the twentieth century.

Partial results have been mainly of two types: theorems saying that every sufficiently large even number is the sum of at most k primes, where proving that one can take $k = 2$ is the ultimate goal, and theorems saying that every sufficiently large even number is the sum of two "nearprimes," that is, numbers with a limited number of prime factors.

The best-known result in the first direction is the 1937 proof by the Russian Vinogradov that every sufficiently large positive integer is the sum of at most four primes. The words "sufficiently large" deserve more explanation. What is meant is that there exists an integer N such that if n is any even integer with $n > N$, then n can be expressed as the sum of four primes. Thus any even integers not the sum of four primes lie in some known, bounded interval, and there is the possibility of checking the even numbers in this interval by direct computation. In Vinogradov's original paper, the value of N was about

$$10^{10^{10^{17.86}}},$$

a number large enough to discourage even a computer, but later mathematicians were able to prove the theorem for smaller values of N.

Actually Vinogradov proved that every sufficiently large *odd* integer is the sum of at most three primes. The result for even integers is an easy consequence of this, since if n is even, then $n - 3$ is odd. Then from $n - 3 = p_1 + p_2 + p_3$, with p_1, p_2, and p_3 prime, follows $n = 3 + p_1 + p_2 + p_2$, and so n is the sum of four primes. Vinogradov's theorem had been preceded by the 1930 proof by Schnirelmann that an integer k exists such that every integer can be written as the sum of at most k primes, but Schnirelmann's value of k was quite large.

Goldbach-type theorems involving the sum of two "nearprimes" go back to Viggo Brun, who proved in 1919 that every sufficiently large even number can be written as the sum of two integers, each of which is the product of at most nine primes. (In these results a prime is counted in a product according to its exponent, so 12 would be considered a product of three primes: 2, 2, and 3.) By 1940 A. Buchstab had reduced the number 9 to 4. In 1948 A. Rényi demonstrated the existence of a number k such that every large even number is the sum of a prime and an integer having at most k prime factors, and by 1963 the mathematician Pan had reduced k to 4. This improved Selberg's 1950 result that every large even number could be expressed as $P + Q$, where PQ had at most five prime factors. In 1965 Buchstab proved the Rényi theorem with $k = 3$. The best theorem to date along these lines is that of the Chinese mathematician Chen, announced in 1966 and published in 1973, that every sufficiently large even integer is the sum of a prime and a number that is the product of at most two primes. If 2 could be replaced by 1 the Goldbach conjecture would be essentially proved.

This little history indicates the international nature of mathematics. A conjecture made by a German who lived in Russia in a letter to a Swiss and published by an Englishman has been worked on by mathematicians of many nations, with the last word so far belonging to a Chinese.

Probably no area of mathematics has been studied more for the sheer enjoyment of its ideas and problems than the theory of numbers. The author hopes that *you*, the reader, will discover some of the beauty and ingenuity that past generations of number theorists, amateur and professional, have found in the subject.

Problems for Chapter 0.

1. List all primes $p < 50$ such that $p + 2$ is also prime.
2. Write each even integer n, $10 < n < 30$, as a sum $p + q$ of two primes with p as small as possible.
3. For which integers n, $6 < n < 14$, is $n^2 + 1$ prime?
4. For which integers n, $1 < n < 14$, is $n^3 + 1$ prime?
5. Prove that if $n > 0$ and $n^2 + 1$ is prime, then $n = 1$ or n is even.
6. Prove that if $n > 0$ and $n^3 + 1$ is prime, then $n = 1$.
7. What is the smallest prime number > 31 of the form $2^n - 1$?
8. What is the smallest prime number > 17 of the form $2^n + 1$?
9. What is the smallest perfect number > 6?
10. For which integers n, $7 < n < 15$, does the decimal expansion of $1/n$ repeat every $n - 1$ digits, but no smaller number of digits?
11. Show that if n is any odd positive integer, and $m = (n^2 - 1)/2$, then $m^2 + n^2 = (m + 1)^2$.
12. Show that if u and v are any integers with $0 < v < u$, and if $x = 2uv$, $y = u^2 - v^2$, and $z = u^2 + v^2$, then $x^2 + y^2 = z^2$.
13. List all primes < 100 of the form $6n + 1$.
14. List all primes < 100 of the form $7n + 3$.
15. List all primes < 100 of the form $8n + 6$.

16. Write each n, $1 < n < 20$, in the form $w^2 + x^2 + y^2 + z^2$, where $0 \leq w \leq x \leq y \leq z$.
17. What is the smallest integer n such that $n = w^2 + x^2 + y^2 + z^2$, $0 \leq w \leq x \leq y \leq z$, has more than one solution?
18. What is the smallest integer n such that $n = w^2 + x^2 + y^2 + z^2$, $0 < w \leq x \leq y \leq z$, has more than one solution?

Euclid

Lived about 300 B.C.

The Greek mathematician Euclid was the most successful textbook writer of all time, his *Elements* having been used in schools for more than 2000 years. Although this work of 13 volumes is usually thought of as being about geometry, it contains much number theory as well, as we will see. Most of the content of the *Elements* was created by other mathematicians, but Euclid organized the material in a way that quickly made his book preeminent.

Little is known about the author himself, and in fact he was often confused with another Euclid (Euclid of Megara), portraits of the wrong man even having been published with editions of the *Elements*. Our Euclid wrote on many subjects besides mathematics, including music, astronomy, mechanics, and optics. (He thought that rays traveled from a person's eyes to the object of his sight.)

The Euclidean algorithm was probably known before Euclid, but its first recorded appearance in print is in Book VII of the *Elements*, as follows: "Two unequal numbers being set out, and the less being continually subtracted from the greater, if the number which is left never measures the one before it until a unit is left, the original numbers will be prime to one another." Here instead of subtracting q times a from b all at once, we successively subtract a until a number less than a remains. Then this remainder is subtracted from a, etc. Euclid says that if the last nonzero remainder is 1, then a and b are relatively prime.

EUCLID

Chapter 1

Divisibility

1.1 The Greatest Common Divisor and Least Common Multiple

The sum, difference, and product of two integers is always an integer, but the quotient may not be. We will start by investigating this most troublesome operation: division of integers.

DEFINITION. divides, divisor, multiple

We say a **divides** b and write $a \mid b$ in case b/a is an integer. In this case, we also say that a is a **divisor** or b and that b is a **multiple** of a. If $a \mid b$ is false, we write $a \nmid b$.

Examples.

$$3 \mid 6, \quad -3 \mid 6, \quad 2 \mid -6, \quad 3 \mid 0, \quad 5 \mid 5, \quad 2 \nmid 5, \quad 8 \nmid -4.$$

◇

DEFINITION. greatest common divisor

We say d is the **greatest common divisor** (gcd) of a and b in case d is the largest of all the integers dividing both a and b, and we write $d = (a, b)$.

Example. Let $a = 4$ and $b = 6$.

Divisors of 4: $1, -1, 2, -2, 4, -4$.
Divisors of 6: $1, -1, 2, -2, 3, -3, 6, -6$.
Common divisors of 4 and 6: $1, -1, 2, -2$.

Thus the greatest common divisor of 4 and 6 is $2 = (4, 6)$. Likewise (since the divisors are the same) $(-4, 6) = (4, -6) = (-4, -6) = 2$. ◇

DEFINITION. least common multiple

We say m is the **least common multiple** (lcm) of a and b in case m is the smallest of all the positive integers that are multiples of both a and b. We write $m = [a, b]$.

Example. Let $a = 4$ and $b = 6$.

Positive multiples of 4: 4, 8, 12, 16, 20, 24, 28, ...
Positive multiples of 6: 6, 12, 18, 24, 30, ...
Common positive multiples of 4 and 6: 12, 24, ...

Thus the least common multiple of 4 and 6 is $12 = [4, 6]$. Likewise (since the multiples are the same) $[-4, 6] = [4, -6] = [-4, -6] = 12$. ♢

The concept of least common multiple should be familiar from arithmetic. To add 3/4 and 1/6 we write

$$\frac{3}{4} + \frac{1}{6} = \frac{9}{12} + \frac{2}{12} = \frac{11}{12}.$$

The 4ths and 6th are converted to 12ths, 12 being the least common multiple of 4 and 6.

The following statements are nearly obvious from the definitions and will be used freely in what follows.

- The following are equivalent: $a \mid b$, $-a \mid b$, $a \mid -b$, $-a \mid -b$.
- If (a, b) is defined, then $(a, b) = (b, a) = (-a, b) = (a, -b) = (-a, -b)$.
- If $[a, b]$ is defined, then $[a, b] = [b, a] = [-a, b] = [a, -b] = [-a, -b]$.

Some Simple Divisibility Theorems

The proofs of Theorems 1.1 and 1.2 below follow quite directly from the definitions of dividing, and the reader may want to try devising his or her own proofs before reading the ones given.

Theorem 1.1. *Given integers a, b, and c, if $a \mid b$ and $b \mid c$, then $a \mid c$.*

Proof. By the definition of $a \mid b$, we know that b/a is an integer. Likewise c/b is an integer. Then so is the product of these quotients an integer, namely,

$$\frac{b}{a} \cdot \frac{c}{b} = \frac{c}{a}.$$

This means that $a \mid c$. □

Note Many authors define $a \mid b$ to mean that $b = ac$ for some integer c. This definition is sometimes easier to use than the one given at the start of this section. The two definitions are equivalent except for the uninteresting case $a = b = 0$. (The relation $0 \mid 0$ is false using our first definition but true for the alternative one.) We will use whichever definition is more convenient in the future. The proof given for the next theorem uses the second definition.

Theorem 1.2. *Given integers a, b, and c, if $a \mid b$ and $a \mid c$, then $a \mid bx + cy$ for any integers x and y.*

Proof. Since $a \mid b$ there exists an integer r such that $b = ra$. Likewise there exists s such that $c = sa$. Then

$$bx + cy = (ra)x + (sa)y = a(rx + sy).$$

This means that $a \mid bx + cy$. □

Note that in particular if $a \mid b$ and $a \mid c$, then by the last theorem $a \mid b + c$ and $a \mid b - c$.

A Property of the Least Common Multiple

Recall that $[a, b]$ is defined to be the smallest positive common multiple of a and b. Let us compute $[15, 10]$ from the definition.

Positive multiples of 15: 15, 30, 45, 60, 75, 90, 105, 120, 135, 150, 165, ...
Positive multiples of 10: 10, 20, 30, 40, 50, 60, 70, 80, 90, 100, 110, 120, ...
Common positive multiples of 15 and 10: 30, 60, 90, 120, ...

We see that $[15, 10] = 30$, the smallest element in the third list. Notice that the other common positive multiples of 15 and 10 appear to be exactly the positive multiples of 30, the least common multiple. This suggests the following result.

Theorem 1.3. *Given integers $a \neq 0$, $b \neq 0$, and m, if $a \mid m$ and $b \mid m$, then $[a, b] \mid m$.*

Proof. It is not quite as easy to see how to form a proof here as in the first two theorems. We wish to show that $m/[a, b]$ is an integer, but it is not clear how this follows from the fact that m/a and m/b are integers.

We will try an indirect proof (see Chapter 0). That is, we will assume that $m/[a, b]$ is not an integer, and try to derive a contradiction from this.

Certainly if $m/[a, b]$ is not an integer, then for some integer q and real number ε

$$\frac{m}{[a, b]} = q + \varepsilon, \qquad \text{where } 0 < \varepsilon < 1.$$

Here q is simply the largest integer less than $m/[a, b]$. (We use the Greek letter ε since italic letters are reserved for integers.) Multiplying through by $[a, b]$ gives

$$m = q[a, b] + \varepsilon[a, b], \qquad \text{and} \qquad 0 < \varepsilon[a, b] < [a, b].$$

(Multiplying the inequalities is justified, since by definition $[a, b] > 0$.)

Notice that $\varepsilon[a, b]$ is the integer $m - q[a, b]$. Let us denote it by r. Then

$$m = q[a, b] + r, \qquad \text{and} \qquad 0 < r < [a, b].$$

Now $r = m - q[a,b]$. By hypothesis $a \mid m$, and $a \mid [a,b]$ by the definition of the lcm. Thus $a \mid r$ by Theorem 1.2 (with $x = 1$ and $y = -q$). The same argument shows that $b \mid r$. We see that r is a common multiple of a and b. But the inequality $0 < r < [a,b]$ contradicts the fact that $[a,b]$ is the *least* positive common multiple of a and b.

Since the assumption that $[a,b]$ does not divide m leads to a contradiction, we conclude that $[a,b]$ divides m. □

Writing Proofs

The proofs of three theorems are given in this section. From them we may draw some simple lessons about how proofs are expressed.

A proof should be written in sentences. These sentences are written in the English language, even though some technical words or symbols may be used. They should start with capital letters (so a sentence should not begin with a symbol) and end with periods. Other conventional punctuation should be used as appropriate.

The most important thing to think about when writing up a proof is who is going to read it. Ivan Niven said, "a proof is something that convinces somebody." It is a communication between humans. It is not sufficient that the writer of the proof believes it; enough detail must be given so that the reader will also understand and believe. The burden of making oneself understood rests on the writer of the proof.

Problems for Section 1.1.

A

1. True or false? $7 \mid 203$, $16 \mid -1000$, $-6 \mid 3$.
2. True or false? $75 \mid 3000$, $71 \mid 0$, $0 \mid -12$.
3. Find $d > 0$ such that $d \mid 18$, $d \nmid 12$, and $36/d \nmid 10$.
4. Find $d > 0$ such that $d \nmid 1000$, $5 \mid d$, $d \mid 60$, and $d/2 \mid 75$.
5. Find $(51, 34)$ and $[51, 34]$.
6. Find $(16, 81)$ and $[16, 81]$.
7. Find all $d > 0$ such that $18 \mid d$ and $d \mid 216$.
8. Find all $d > 0$ such that $20 \mid d$ and $d \mid 300$.
9. What are all the divisors of 24?
10. What are all the divisors of 30?
11. What are the multiples of 4 between -25 and 25?
12. What are the multiples of 5 between -42 and 42?
13. Make a table showing b, (a,b), $[a,b]$, and $(a,b)[a,b]$ for $a = 8$ and b running from 1 to 9.
14. Make a table as in the problem 13 for $a = 9$ and b running from 1 to 10.

B

15. For what integers a is $1 \mid a$ true?
16. For what integers a is $a \mid 0$ true?
17. For what integers a is $a \mid b$ true for all integers b?

18. Why does the word "positive" appear in the definition of $[a, b]$ but not in the definition of (a, b)?

True-False. In the next nine problems, tell which statements are true and give counterexamples to those that are false. Assume a, b, and c are arbitrary nonzero integers.

19. If $ab > 0$, then $[a, b] \leq ab$.
20. If $c \mid a$ and $c \mid b$, then $[a, b] \leq ab/c$.
21. If $(a, b) = 1$ and $(a, c) = 1$, then $(b, c) = 1$.
22. $(a, b) \mid [a, b]$.
23. If $b \mid c$, then $(a, b) \leq (a, c)$.
24. If $b \mid c$, then $[a, b] \leq [a, c]$.
25. If $a \mid b$ and $b \mid c$, then $a \mid c$.
26. $(ac, bc) = c(a, b)$.
27. $(ac, bc) = |c|(a, b)$.

C

28. Show that (a, b) is defined if and only if a and b are not both 0.
29. Show that $[a, b]$ is defined if and only if neither a nor b is 0.
30. Show that neither (a, b) nor $[a, b]$ can be negative.
31. Show why $ab/[a, b]$ must be an integer whenever $[a, b]$ is defined.
32. Show that a and $-a$ have the same divisors and the same multiples.
33. Show that if (a, b) is defined, then it equals $(|a|, |b|)$, and that if $[a, b]$ is defined, then it equals $[|a|, |b|]$.
34. Prove that if $a \mid b$ by our original definition, then there exists an integer c such that $b = ac$.
35. Prove that if a and b are not both 0 and if there exists an integer c such that $b = ac$, then $a \mid b$ by our original definition.
36. Show that if a, b, and $d > 0$ are integers such that $d \mid a$, $d \mid b$, and $(a/d, b/d) = 1$, then $d = (a, b)$.

1.2 The Division Algorithm

Turning Tricks into Methods

Our proof of the last theorem involved an idea that will be useful in the future. Let us try to write it down in a general form, as a theorem that may simply be quoted if the same trick is needed again. (Someone has said that in mathematics if a "trick" is used twice it becomes a "method.")

The idea occurred near the end of the proof when we expressed m as a multiple of $[a, b]$ plus an integer r, where $0 \leq r < [a, b]$. (In the proof of the theorem the assumption that $[a, b]$ didn't divide m implied that $r > 0$.) This is really a familiar arithmetic fact. For example, one knows without doing the actual computation that if 592 is divided by 7 the answer is some integer q plus $r/7$, where $r = 0, 1, 2, 3, 4, 5$, or 6. Thus (multiplying by 7)

$$592 = 7q + r,\ 0 \leq r < 7.$$

In fact, if you and I both do this computation, we should expect to get the same values for q and r. In other words, the **quotient** q and **remainder** r are uniquely determined.

Theorem 1.4 (the division algorithm). *Suppose a and b are integers, $a > 0$. Then there exist unique integers q and r, $0 \leq r < a$, such that $b = aq + r$.*

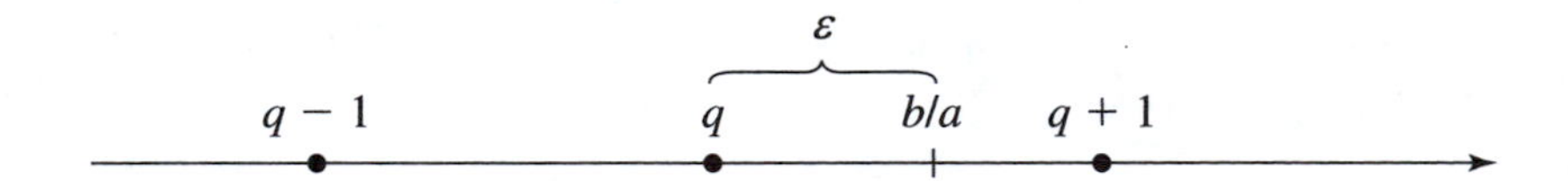

Proof. First we show q and r exist. Suppose the real number b/a is $q + \varepsilon$, where q is an integer and $0 \leq \varepsilon < 1$. Then

$$b = a(q + \varepsilon) = aq + a\varepsilon.$$

From $b = aq + a\varepsilon$ if follows that $a\varepsilon$ is an integer, and $0 \leq a\varepsilon < a$. (Why?) Set $a\varepsilon = r$. Thus $b = aq + r$, and $0 \leq r < a$.

Next we show q and r are unique. Suppose

$$\begin{aligned} b &= aq + r, & 0 \leq r < a, \\ b &= aq' + r', & 0 \leq r' < a. \end{aligned}$$

Subtracting the two equations and transposing $r - r'$ gives

$$r' - r = a(q - q').$$

But adding the inequalities $-a < -r \leq 0$ and $0 \leq r' < a$ and dividing by a yields

$$-1 < \frac{r' - r}{a} < 1.$$

Since $(r' - r)/a$ is the integer $q - q'$, we see $q - q' = 0$. This implies $q = q'$ and $r = r'$. $\square$

Examples. If $a = 7$ and $b = 592$, then $q = 84$ and $r = 4$, since

$$592 = 7 \cdot 84 + 4 \quad \text{and} \quad 0 \leq 4 < 7.$$

If $a = 54$ and $b = 115$, then $q = 23$ and $r = 0$, since

$$115 = 5 \cdot 23 + 0 \quad \text{and} \quad 0 \leq 0 < 54.$$

If $a = 9$ and $b = 4$, then $q = 0$ and $r = 4$, since

$$4 = 9 \cdot 0 + 4 \quad \text{and} \quad 0 \leq 4 < 9.$$

If $a = 6$ and $b = -22$, then $q = -4$ and $r = 2$, since

$$-22 = 6(-4) + 2 \quad \text{and} \quad 0 \leq 2 < 6.$$

The Division Algorithm with a Calculator

If the numbers a and b are large it may save time to use a calculator to compute the q and r of the division algorithm. As we saw in its proof, q is just the largest integer less than or equal to b/a. If $a = 375$ and $b = 49{,}162$, one calculator gives $b/a = 131.0986667$, and so $q = 131$.

Now $r = a(b/a - q)$. Thus in our example $r = 375(131.098667 - 131) = 37.0000125$ on the same calculator. Of course r is the integer 37, and the .0000125 is due to round-off error.

In general to find q and r we:

1. Compute b/a; q is the largest integer not exceeding this number.
2. Subtract q from b/a.
3. Multiply by a to get r.

Looking for a Pattern

Number theory has the advantage that we may sometimes find theorems simply by experimentation. We will try to find some nonobvious relation or formula involving (a, b) or $[a, b]$. Let us see what happens when we combine (a, b) and $[a, b]$ in various ways. To keep things simple we will set $a = 6$ and let b vary.

a	b	(a,b)	$[a,b]$	$(a,b)+[a,b]$	$[a,b]-(a,b)$	$(a,b)[a,b]$
6	1	1	6	7	5	6
6	2	2	6	8	4	12
6	3	3	6	9	3	18
6	4	2	12	14	10	24
6	5	1	30	31	29	30
6	6	6	6	12	0	36
6	7	1	42	43	41	42

The only column showing much regularity is the one giving $(a, b)[a, b]$. Perhaps we should try a different value of a, say $a = 4$.

a	b	(a,b)	$[a,b]$	$(a,b)[a,b]$
4	1	1	4	4
4	2	2	4	8
4	3	1	12	12
4	4	4	4	16
4	5	1	20	20
4	6	2	12	24

It appears that perhaps $(a, b)[a, b] = ab$ for all integers a and b. We must be careful, however; what if a or b were negative? It is easy to see from the definitions that (a, b) and $[a, b]$ are always positive, but ab might not be. We will avoid this problem by assuming that a and b are both positive.

Theorem 1.5. *If a and b are positive integers, then $(a, b)[a, b] = ab$.*

Proof. In this proof we will leave out the reasons for several steps in the argument, asking the reader to supply them. The correct reasons will be found at the end of the problems for this section.

The equation we are trying to prove may be rewritten as

$$[a, b] = \frac{ab}{(a, b)}.$$

One way to prove this would be to show that the expression on the right satisfies the definition of $[a, b]$. It is easy to see that it is positive. [**Question 1: Why?**] Also $ab/(a, b)$ is an integer since it is a times $b/(a, b)$, and the latter is an integer. [**Question 2: Why?**]

Is $ab/(a, b)$ a common multiple of a and b? We have

$$\frac{ab/(a, b)}{a} = \frac{b}{(a, b)},$$

which we have just seen is an integer, so $ab/(a, b)$ is a multiple of a. In the same way we could prove that $ab/(a, b)$ is a multiple of b. Thus $ab/(a, b)$ is a common multiple of a and b, from which we conclude that

$$\frac{ab}{(a,b)} \geq [a,b].$$ **[Question 3: Why?]**

If we could prove this inequality in the opposite direction, we would be done. Let us start over by considering $ab/[a,b]$. This is an integer. **[Question 4: Why?]** Also

$$\frac{a}{ab/[a,b]} = \frac{[a,b]}{b},$$

which is an integer. **[Question 5: Why?]** Thus $ab/[a,b]$ divides a. A similar proof shows that $ab/[a,b]$ divides b. We see that $ab/[a,b]$ is a common divisor of a and b. Thus

$$\frac{ab}{[a,b]} \leq (a,b).$$ **[Question 6: Why?]**

Since the two inequalities we have proved are equivalent to

$$ab \geq (a,b)[a,b] \qquad \text{and} \qquad ab \leq (a,b)[a,b],$$

the theorem follows. □

Problems for Section 1.2.

A

Find the q and r guaranteed by the division algorithm for each pair a, b in problems 1 through 12.

1. $a = 13$, $b = 380$
2. $a = 15$, $b = 421$
3. $a = 720$, $b = 155$
4. $a = 339$, $b = 17$
5. $a = 17$, $b = 51$
6. $a = 21$, $b = 105$
7. $a = 19$, $b = 0$
8. $a = 35$, $b = 0$
9. $a = 7$, $b = -30$
10. $a = 9$, $b = -29$
11. $a = 43$, $b = -500$
12. $a = 47$, $b = -500$

B

13. What are all the common divisors of 12 and 18?
14. What are all the common divisors of 45 and 75?
15. What are all the common multiples of 4 and 6?
16. What are all the common multiples of 27 and 18?

True-False. In the next eight problems, tell which statements are true and give counterexamples for those that are false. Assume a, b, c, and d are arbitrary integers with $a > 0$ and c and d nonzero.

17. There exist integers q and r, $0 \leq r < c$, such that $b = cq + r$.
18. There exist integers q and r, $0 \leq r < |c|$, such that $b = cq + r$.
19. There exist integers q and r, $|r| \leq a/2$, such that $b = aq + r$.
20. There exist integers q and r, $|r| < a/2$, such that $b = aq + r$.

21. The set of common divisors of b and c is the set of divisors of (b, c).
22. The set of common multiples of c and d is the set of multiples of $[c, d]$.
23. If b is a divisor of c, and $b > (c, d)$, then b is not a divisor of d.
24. If b is a multiple of c, and $b < [c, d]$, then b is not a multiple of d.

The next three problems refer to the following conditions: (i) $a \mid b$, (ii) $a2^a \mid b^2$, (iii) $2^a \leq b$.

25. Give positive integers a and b such that (i) and (ii) hold, but not (iii).
26. Give positive integers a and b such that (i) and (iii) hold, but not (ii).
27. Give positive integers a and b such that (ii) and (iii) hold, but not (i).

C

28. Prove that if $ab \neq 0$, then $(a, b)[a, b] = |ab|$.
29. Prove that $(a, a + 2)$ is 2 is a is even and 1 if a is odd.
30. Prove that if $a > 0$, then $[a, a + 2] = a(a + 2)/2$ if a is even and $a(a + 2)$ if a is odd.
31. Prove that if $x > 0$, then $(a, a + x)$ is (a, x).
32. Prove that if $d \mid a$, $d \mid b$, and $d \mid c$, and if x, y, and z are any integers, then d divides $ax + by + cz$.
33. Prove that if $a \mid b$, $b \mid c$, and $c \mid d$, then $a \mid d$.
34. Prove that if $a \mid b$ and $b \mid c$, then a divides $ax + by + cz$ for any integers x, y, and z.
35. Prove that if $b \neq 0$ and $a = bx + cy$, then $(b, c) \leq (a, b)$.
36. Prove that with the hypotheses of the last problem, $(b, c) \mid (a, b)$.

Answers to questions in the proof of the Theorem 1.5.

1. By hypothesis $ab > 0$, and $(a, b) > 0$ since if d is a common divisor, then so is $|d|$.

2. By the definition of (a, b).

3. Since $[a, b]$ is the *least* common multiple of a and b.

4. By Theorem 1.3.

5. By the definition of $[a, b]$.

6. Since (a, b) is the *greatest* common divisor of a and b.

1.3 The Euclidean Algorithm

Calculating the Greatest Common Divisor

Although we have proved various theorems about the greatest common divisor (a, b), we have no easy way to compute it. Of course, we could simply list all the divisors of the smaller of a and b, say a, and look for the largest of these that also divides b. If a is large, however, this could be tedious. It turns out that the division algorithm yields a way to make the task easier.

Let us denote (a, b) by d, and suppose $0 < a < b$. We write

$$b = aq + r, \ \ 0 \leq r < a.$$

Then $r = b - aq$, and since d divides both a and b, Theorem 1.2 implies that d divides r also. Since $r < a$ this fact may make finding d easier.

In fact since d divides both r and a, $d \leq (r, a)$ by the definition of the latter. But applying Theorem 1.2 to $b = aq + r$ shows, in the same way, that (a, r) divides b. Since (a, r) divides a by definition, we have $(a, r) \leq (a, b) = d$ by the definition of (a, b).

We see from the above paragraph that $(a, b) = (r, a)$. Actually nothing more that the equation $b = aq + r$ was used in the proof, and the fact that q and r came from the division algorithm was irrelevant. We have proved the following:

Lemma 1.6. *For any integers $a > 0$, b, c, and k, if $a = bk + c$, then $(a, b) = (b, c)$.*

This fact is especially useful when applied to the equation of the division algorithm as above because $a > r$ and $b > a$, so we have replaced a and b by a smaller pair. In fact the process can be repeated until the gcd is obvious.

Example. Suppose we want $(504, 123)$. Set $a = 123$ and $b = 504$. Now

$$504 = 123 \cdot 4 + 12.$$

Thus $r = 12$ and $(123, 504) = (12, 123)$. Likewise

$$123 = 12 \cdot 10 + 3,$$

so $(12, 123) = (3, 12)$.

Although clearly $(3, 12) = 3$, let us see how far we can carry this process. Now

$$12 = 3 \cdot 4 + 0,$$

and so $(3, 12) = (0, 3)$. This is clearly 3 since any positive number divides 0. Thus $(504, 123) = 3$. ◇

Theorem 1.7 (the Euclidean algorithm). *Let a and b be integers, $a > 0$. Apply the division algorithm repeatedly as follows:*

$$\begin{aligned}
b &= aq_1 + r_1, & 0 &< r_1 < a,\\
a &= r_1 q_2 + r_2, & 0 &< r_2 < r_1,\\
r_1 &= r_2 q_3 + r_3, & 0 &< r_3 < r_2,\\
&\vdots & &\vdots\\
r_{n-2} &= r_{n-1} q_n + r_n, & 0 &< r_n < r_{n-1},\\
r_{n-1} &= r_n q_{n+1}.
\end{aligned}$$

Let r_n be the last nonzero remainder. Then $(a, b) = r_n$.

Proof. The process must terminate, since the remainders form a strictly decreasing sequence of nonnegative integers. Then, as we have seen,

$$(a, b) = (a, r_1) = (r_1, r_2) = \cdots = (r_{n-1}, r_n) = (r_n, 0) = r_n.$$

□

Example. We will compute $(158, 188)$, using arrows to emphasize the pattern.

$$\begin{aligned}
188 &= 158 \cdot 1 + 30\\
158 &= 30 \cdot 5 + 8\\
30 &= 8 \cdot 3 + 6\\
8 &= 6 \cdot 1 + 2\\
6 &= 2 \cdot 3 + 0
\end{aligned}$$

Thus $(158, 188) = 2$. ◇

Solving the Euclidean Algorithm Equations Backward

It is possible to use the Euclidean algorithm equations to write (a, b) in the form $ax + by$. We start with the equation having (a, b) as a remainder and work backwards. We will do this for the above example.

$$\begin{aligned}
2 &= 8 - 6 \cdot 1 & &\\
&= 8 - (30 - 8 \cdot 3)1 & &= -30 + 8 \cdot 4\\
&= -30 + (158 - 30 \cdot 5)4 & &= 158 \cdot 4 - 30 \cdot 21\\
&= 158 \cdot 4 - (188 - 158)21 & &= -188 \cdot 21 + 158 \cdot 25
\end{aligned}$$

Thus $(158, 188) = 158x + 188y$ for $x = 25$ and $y = -21$.

If either a or b is negative, we simply deal with $|a|$ and $|b|$ and insert the correct signs at the end of the problem. For example, suppose we want to find x and y such that $ax + by = (a, b)$ for $a = -56$ and $b = -21$. We work with 21 and 56.

$$\begin{aligned} 56 &= 21 \cdot 2 + 14 \\ 21 &= 14 \cdot 1 + 7 \\ 14 &= 7 \cdot 2 + 0 \end{aligned}$$

Thus $7 = (21, 56) = (-21, -56)$. Also

$$\begin{aligned} 7 &= 21 - 14 \cdot 1 \\ &= 21 - (56 - 21 \cdot 2)1 \quad = \quad -56 + 21 \cdot 3 \\ &= (-56)(1) + (-21)(-3). \end{aligned}$$

We see $x = 1$ and $y = -3$.

The proof that this can always be done will be left for the problems at the end of this section.

Theorem 1.8. *If a and b are integers such that (a, b) is defined then there exist integers x and y such that $(a, b) = ax + by$.*

DEFINITION. linear combination

We say k is a **linear combination** of a and b in case there exist integers x and y such that $k = ax + by$.

The Duality of the gcd and lcm

The reader may have noticed a certain duality between the concepts of least common multiple and greatest common divisor. Replacing "least" with "greatest" and "multiple" with "divisor" changes the definition of the former into that of the latter. New (possible) theorems can be created in a like way. For example, the following theorem restates Theorem 1.3 as its part (1); part (2) is the dual result.

Theorem 1.9. *Let a, b, m, and d be integers.*

1. *If m is any common multiple of a and b, then $[a, b]$ divides m.*

2. *If d is any common divisor of a and b, not both of which are 0, then d divides (a, b).*

Proof. Only (2) needs to be proved. By the last theorem there exist x and y such that $(a, b) = ax + by$, which is divisible by d by Theorem 1.2. □

Problems for Section 1.3.

A

Find (a, b) for each pair given in problems 1 through 8 using the Euclidean algorithm. Then solve the equations backward to find x and y such that $ax + by = (a, b)$.

1. $a = 217$, $b = 341$
2. $a = 117$, $b = 247$
3. $a = 143$, $b = 451$
4. $a = 165$, $b = 465$
5. $a = 89$, $b = 55$
6. $a = 123$, $b = 76$
7. $a = -899$, $b = 2030$
8. $a = -4050$, $b = -1728$
9. Find x and y such that $26x + 14y = (26, 14)$ with x positive but as small as possible.
10. Find x and y such that $27x + 15y = (27, 15)$ with x positive but as small as possible.
11. Use problem 1 to solve $217x + 341y = 62$.
12. Use problem 2 to solve $117x + 247y = 39$.
13. Find all solutions to $2x + 3y = 50$ in positive integers.
14. Find all solutions to $3x + 4y = 60$ in positive integers.
15. Find all solutions to $4x + 6y = 60$ in positive integers.
16. Find all solutions to $6x + 9y = 91$ in positive integers.

B

17. Why is $4x + 6y = 25$ unsolvable?
18. Why is $361x + 2109y = 1000$ unsolvable?
19. Determine all integers x such that $2x + 3y = 1$ is solvable.
20. Determine all integers x such that $3x + 2y = 4$ is solvable.

C

21. Prove that if r'' is a linear combination of r and r', and if r and r' are each linear combinations of a and b, then r'' is a linear combination of a and b.
22. Use the previous problem and the Euclidean algorithm to prove Theorem 1.8.
23. Let a and b be integers, not both 0, and let S be the set of all linear combinations of a and b. Show that S contains at least one positive element.
24. Let c be the least positive element of the set S of the previous problem. Show that $c \mid a$ and $c \mid b$. (**Hint:** Use the division algorithm.)
25. Use the previous problem to provide another proof of Theorem 1.8 of this section. (**Hint:** Prove that $c = (a, b)$.)
26. Use Theorem 1.8 to prove Theorem 1.10 in the next section.
27. Why is the assumption that a and b are not both 0 needed in part (2) of Theorem 1.9 but not in part (1)?
28. Prove that if a and b are positive integers, then there exist integers x and y such that $1/[a, b] = x/a + y/b$.
29. Prove that if $(a, b) = 1$ and $c \neq 0$, then $(ac, bc) \mid c$.

1.4 Linear Combinations

When $ax + by = c$ Is Solvable

We can find x and y such that $ax + by = c$ not only when $c = (a, b)$, but also whenever c is any multiple of (a, b). For if $ax + by = (a, b)$ and $c = k(a, b)$, then $a(kx) + b(ky) = c$. Conversely, if $ax + by = c$, then (a, b) must divide c by Theorem 1.2.

Theorem 1.10. *Given integers a, b, and c with a and b not both 0, there exist integers x and y such that $ax + by = c$ if and only if $(a, b) \mid c$.*

Corollary 1.11. *Let a and b be integers. There exist integers x and y such that $ax + by = 1$ if and only if $(a, b) = 1$.*

Corollary 1.12. *Let a, a', and b be integers. If $(a, b) = 1$ and $(a', b) = 1$, then $(aa', b) = 1$.*

Proof. By the preceding corollary there exist integers x, y, x', and y' such that

$$ax + by = 1 \quad \text{and} \quad a'x' + by' = 1.$$

Multiplying these equations gives

$$(aa')(xx') + b(axy' + a'x'y + byy') = 1.$$

Thus (again by the first corollary) $(aa', b) = 1$. □

DEFINITION. relatively prime

We say a and b are **relatively prime** in case $(a, b) = 1$. In a certain sense if a and b are relatively prime, then a and b "have nothing to do with each other." The next theorem illustrates this idea.

Theorem 1.13. *If a, b, and c are integers such that $(a, b) = 1$ and $a \mid bc$, then $a \mid c$.*

Proof. By the first corollary, there exist x and y such that $ax + by = 1$. Then $acx + bcy = c$. Since $a \mid a$ and $a \mid bc$ we apply Theorem 1.2 to conclude $a \mid c$. □

WARNING One of the most common mistakes made by beginning number theory students is to assume that if $a \mid bc$, then $a \mid b$ or $a \mid c$. This is not generally true, and the reader should construct a counterexample before proceeding.

If we remove the "common part" (a, b) from a and b, we would expect the remaining integers to have nothing left in common. The following theorem expresses this idea precisely.

Theorem 1.14. *If a and b are any integers and $(a, b) = d$, then a/d and b/d are relatively prime.*

Proof. We can write $ax + by = d$. Then $(a/d)x + (b/d)y = 1$, and so a/d and b/d are relatively prime by Corollary 1.11. □

All Solutions to $ax + by = c$

If $(a, b) \mid c$, our Euclidean algorithm method enables us to find one solution of $ax + by = c$. Consider the equation $9x + 24y = 15$, for example. We write

$$\begin{aligned} 24 &= 9 \cdot 2 + 6 \\ 9 &= 6 \cdot 1 + 3 \\ 6 &= 3 \cdot 2 + 0. \end{aligned}$$

Since $(9, 24) = 3$ and $3 \mid 15$ we know a solution exists. In fact

$$\begin{aligned} 3 &= 9 - 6 \cdot 1 \\ &= 9 - (24 - 9 \cdot 2)1 = 9 \cdot 3 - 24 \cdot 1. \end{aligned}$$

Then $15 = 3 \cdot 5 = 9 \cdot 15 - 24 \cdot 5$. We see $x = 15$ and $y = -5$ is a solution. But there are others, for example $x = -1$ and $y = 1$.

We will study how to find all solutions to $ax + by = c$, given a particular one, say x_0, y_0. We assume that a and b are nonzero since finding all solutions is easy otherwise. Suppose

$$ax_0 + by_0 = c \quad \text{and} \quad ax + by = c,$$

so x, y is another solution. Then

$$a(x - x_0) = b(y_0 - y).$$

We see that $a \mid b(y_0 - y)$. If we knew that a and b were relatively prime, then we could conclude by Theorem 1.13 that a divided $y_0 - y$; but of course it is possible that $(a, b) = d > 1$.

We will get around this by using the equation

$$\frac{a}{d}(x - x_0) = \frac{b}{d}(y_0 - y),$$

where a/d and b/d are relatively prime by Theorem 1.14. This equation implies $(a/d) \mid y_0 - y$.

Suppose $y_0 - y = (a/d)t$, where t is some integer. Then our equation yields

$$x - x_0 = \frac{b}{a}(y_0 - y) = \frac{b}{d} \cdot t.$$

Solving for x and y gives

$$x = x_0 + \frac{b}{d} \cdot t \quad \text{and} \quad y = y_0 - \frac{a}{d} \cdot t.$$

Let us review what we have done so far. We assumed x_0, y_0 was a particular solution to $ax + by = c$, and also that x, y was some other solution. This forced us to conclude that

$$x = x_0 + \frac{b}{d} \cdot t, \quad y = y_0 - \frac{a}{d} \cdot t, \tag{1.1}$$

where t is some integer and $d = (a, b)$.

We see any other solution must be of the form (1.1). The question remains whether any pair x, y of the form (1.1) must be a solution to the original equation. Plugging in such a pair shows that this is the case. Indeed

$$a\left(x_0 + \frac{b}{d} \cdot t\right) + b\left(y_0 - \frac{a}{d} \cdot t\right) = ax_0 + by_0 + \frac{abt}{d} - \frac{abt}{d} = c.$$

Theorem 1.15. *Suppose $a \neq 0$, $b \neq 0$, and c are integers. Let x_0, y_0 be a particular solution to $ax + by = c$. Then all solutions are given by*

$$x = x_0 + \frac{b}{d} \cdot t, \quad y = y_0 - \frac{a}{d} \cdot t,$$

as t runs through the integers, where $d = (a, b)$.

Example. We have seen that one solution to $9x + 24y = 154$ is $x_0 = 15$, $y_0 = -5$, and that $(9, 24) = 3$. Then all solutions are

$$x = 15 + \frac{24}{3}t = 15 + 8t, \quad y = -5 - \frac{9}{3}t = -5 - 3t.$$

Taking $t = -2$ gives the solution $x = -1$, $y = 1$ noted previously. ♢

Example. Find all positive integers x, y such that $4x + 6y = 100$.

We note that $(4, 6) = 2$ and $2 \mid 100$, so solutions do exist. In fact $x = 25$ and $y = 0$ is one solution, though not in positive integers.

Taking $a = 4$, $b = 6$, and $d = 2$, the general solution is

$$x = 25 + \frac{6}{2}t = 25 + 3t, \quad y = 0 - \frac{4}{2}t = -2t.$$

In this case, we want $x = 25 + 3t > 0$ and $y = -2t > 0$. The first inequality amounts to $3t > -25$ or $t > -8\frac{1}{3}$, while $-2t > 0$ says $t < 0$. Thus t runs from -1 to -8, and we get the solutions shown below.

t	-1	-2	-3	-4	-5	-6	-7	-8
x	22	19	16	13	10	7	4	1
y	2	4	6	8	10	12	14	16

♢

Example. Find all solutions to $15x + 16y = -1000$ with $x \leq 0$ and $y \leq 0$.

Clearly $(15, 16) = 1$ and

$$15(-1) + 16(1) = 1.$$

Thus $x_0 = 1000$ and $y_0 = -1000$ is one solution to the original equation. The general solution (taking $a = 15$ and $b = 16$) is

$$x = x_0 + bt = 1000 + 16t, \quad y = y_0 - at = -1000 - 15t.$$

Also $x = 1000 + 16t \leq 0$ says

$$16t \leq -1000 \quad \text{or} \quad t \leq -62\tfrac{1}{2},$$

and $y = -1000 - 15t \leq 0$ says

$$15t \geq -1000 \quad \text{or} \quad t \geq -66\tfrac{2}{3}.$$

Since t is an integer we see $t = -63, -64, -65$, or -66.

$-66\frac{2}{3}$ -66 -65 -64 -63 $-62\frac{1}{2}$ -62

t	-63	-64	-65	-66
x	-8	-24	-40	-56
y	-55	-40	-25	-10

◊

Example. A man paid \$11.37 for some 39-cent pens and 69-cent pens. How many of each did he buy?

Suppose he bought x pens at 39 cents and y pens at 69 cents. Then we wish to solve

$$39x + 69y = 1137, \quad x \geq 0,\ y \geq 0.$$

We have $(39, 69) = 3$ and $3 \mid 1137$, so the equation has solutions (although we still don't know if any are nonnegative). Since we found $(39, 69) = 3$ without recourse to the Euclidean algorithm we will divide by 3 and solve the simpler system

$$13x + 23y = 379, \quad x \geq 0,\ y \geq 0.$$

Using the Euclidean algorithm on 13 and 23 and solving the equations backward leads to the equation

$$13(-7) + 23(4) = 1.$$

Thus we have the particular solution $x_0 = 379(-7) = -2653$, $y_0 = 379(4) = 1516$. The general solution (using $a = 13$, $b = 23$, $d = 1$) is

$$x = -2653 + 23t, \quad y = 1516 - 13t.$$

Here $x \geq 0$ and $y \geq 0$ amount to

$$t \geq \frac{2653}{23} = 115\frac{8}{23} \quad \text{and} \quad t \leq \frac{1516}{13} = 116\frac{8}{13}.$$

Thus $t = 116$, $x = -2653 + 23(116) = 15$, and $y = 1516 - 13(116) = 8$. The man bought 15 pens at 39 cents and 8 pens at 69 cents. ◊

Problems for Section 1.4.

A

In problems 1 through 6, tell whether or not the equation has a solution.

1. $3x + 5y = 50{,}001$
2. $6x + 9y = 60{,}001$
3. $21x - 14y = 10{,}000$
4. $-12x + 42y = 366$
5. $529x + 2024y = 391$
6. $851x + 1147y = 481$

Use the Euclidean algorithm method to find one solution to the equations in problems 7 through 12.

7. $7x + 20y = 3$
8. $8x + 21y = 5$
9. $66x + 51y = 300$
10. $65x + 50y = 300$
11. $200x - 300y = 400$
12. $55x + 200y = -100$

In problems 13 through 19, find all solutions with x and y positive.

13. $5x + 6y = 100$
14. $6x + 7y = 200$
15. $6x + 8y = 120$
16. $9x + 6y = 150$
17. $121x + 561y = 13{,}200$
18. $169x + 663y = 2340$
19. $621x + 1026y = 49{,}194$

B

20. If $abc \neq 0$, is it possible for $ax + by = c$ to have infinitely many solutions in positive integers?
21. For what triples a, b, c is it true that for each integer x there is an integer y such that $ax + by = c$?
22. A girl spent $100.64 on posters. Some cost $4.98 and some $5.98. How many did she buy?
23. A man bought three dozen oranges and two dozen apples. His change from a $10 bill was $1.96. One orange costs more than 10 cents and one apple more than 15 cents. How much does an orange cost?
24. A roadside stand bought 11 large baskets of eggs from a farmer and sold 39 small baskets of eggs, which hold fewer than a dozen. There were 19 eggs left over. How many eggs does a large basket hold?
25. Farmer Jones owes Farmer Brown $10. Neither has any cash, but Jones has 14 cows, which he values at $185 each. He suggests paying his debt in cows, with Brown making change by giving Jones some of his pigs at $110 each. Is this possible, and how?
26. Brown wants his $10, but says his pigs are worth $111 each. Is the deal still possible, and how?
27. Jones offers to trade his cows for $184 each if Brown accepts $100 per pig. Now what?
28. Harry bought some 39-cent pens. He paid with a $10 bill, getting 42 cents in coins in change (plus maybe some bills). How many did he buy?
29. The same Harry as in the last problem bought 28 more expensive pens at another store, receiving 44 cents change in coins from a $20 bill, plus perhaps some bills. How much were the pens?

30. A retailer would like to sell big boxes of Christmas cards for less than \$15. He wants to get 34 cents for each card, and wants the price of a box to end in 99—for example, \$7.99. If this is impossible he would like it to end in 98, 97, or 96, etc., in order of decreasing desirability. How many cards should he put in a box?
31. Alan paid \$133.98 for some \$1.29 pens and \$2.49 notebooks. He didn't buy 100 pens and 2 notebooks, even though this adds up to the right amount. How many of each *did* he buy?
32. A woman paid a \$9.75 debt using only dimes and quarters. In how many ways is this possible?
33. Tina spent \$86.66 on some \$1.47 puzzles and on 371 erasers. How much does an eraser cost?
34. A collector has 50 toy planes worth \$2.80 each. He swaps some of them for some \$5.61 toy cars and \$4.42 toy trucks. How many of each toy are involved?
35. Find integers x and y such that $119x - 161y = 4900$ with $|x - y|$ as small as possible.

C

36. Prove that if $a \mid bc$, then $a \mid (a, b)c$.
37. Give a proof of Theorem 1.14 using nothing more than the definition of (a, b).
38. Show that if $(a, b) \mid c$ and if $ab < 0$, then $ax + by = c$ has at least one solution with x and y positive.
39. Prove that if $(a, b) = 1$, then $(a^2, b^2) = 1$.

1.5 Congruences

A Table for (a, b)

At the beginning of this chapter we discovered a relation between (a, b) and $[a, b]$ by looking at a table. We will return to this technique, fixing a at 6 and letting b vary.

a	6	6	6	6	6	6	6	6	6	6	6	6	6
b	1	2	3	4	5	6	7	8	9	10	11	12	13
(a, b)	1	2	3	2	1	6	1	2	3	2	1	6	1

The values of (a, b) seem to repeat the pattern 1–2–3–2–1–6 again and again. It appears that whatever $(6, b)$ is, $(6, b+6)$ will be the same, as will $(6, b+12)$, $(6, b+18)$, etc. For example $(6, 1) = (6, 7) = (6, 13) = 1$. The numbers 1, 7, and 13 have the property that any two of them differ by a multiple of 6. This idea is important enough to have its own symbolism.

DEFINITION. congruence

Let $m > 0$. We say that a is **congruent to** b **modulo** m and write

$$a \equiv b \pmod{m}$$

in case $m \mid b - a$. Such a statement is called a **congruence** and m is called its **modulus**.

Examples. Since $6 \mid 6 = 7 - 1$, $1 \equiv 7 \pmod{6}$.

Since $6 \mid 13 - 1$, $1 \equiv 13 \pmod{6}$.

Since $6 \nmid 5 = 7 - 2$, $2 \not\equiv 7 \pmod{6}$.

Why Congruence?

The reader may think that we do not need a new and slightly longer way to say $m \mid b - a$. However, the notation $a \equiv b \pmod{m}$ is useful because it suggests an equation, and in fact congruences share many properties with equations. A symbolism that emphasizes this similarity is useful for suggesting new results and making them easier to remember.

We will soon see for example that if $a \equiv b \pmod{m}$ and $a' \equiv b' \pmod{m}$, then $aa' \equiv bb' \pmod{m}$. In other words, congruences can be multiplied. Without the new symbolism, however, the result would say that if $m \mid b - a$ and $m \mid b' - a'$, then $m \mid bb' - aa'$, a much more awkward formulation.

The next theorem lists a number of ways in which congruences behave like equations. Readers are encouraged to try to find their own proofs before reading those given. Part (5) is the hardest.

Theorem 1.16. *Let a, a', b, b', c, d, and m be integers, with $d > 0$ and $m > 0$.*

1. *Always* $a \equiv a \pmod{m}$.
2. *If* $a \equiv b \pmod{m}$, *then* $b \equiv a \pmod{m}$.
3. *If* $a \equiv b \pmod{m}$ *and* $b \equiv c \pmod{m}$, *then* $a \equiv c \pmod{m}$.
4. *If* $a \equiv b \pmod{m}$ *and* $a' \equiv b' \pmod{m}$, *then* $a + a' \equiv b + b' \pmod{m}$.
5. *If* $a \equiv b \pmod{m}$ *and* $a' \equiv b' \pmod{m}$, *then* $aa' \equiv bb' \pmod{m}$.
6. *If* $a \equiv b \pmod{m}$ *and* $d \mid m$, *then* $a \equiv b \pmod{d}$.

Proof.

1. Always $m \mid a - a = 0$.
2. If $b - a = mk$, then $a - b = m(-k)$.
3. We have $m \mid b-a$ and $m \mid c-b$, so $m \mid c-a = (c-b)+(b-a)$ by Theorem 1.2.
4. Since $m \mid b-a$ and $m \mid b'-a'$, we have $m \mid (b-a)+(b'-a') = (b+b')-(a+a')$ by Theorem 1.2.
5. Since $m \mid b-a$ and $m \mid b'-a'$, we have $m \mid (b-a)b'+a(b'-a') = bb'-aa'$ by Theorem 1.2.
6. Since $m \mid b-a$ and $d \mid m$, $d \mid b-a$ by Theorem 1.1.

□

Note The proof for part (5) uses the common mathematical trick of adding and subtracting the same thing to get an agreeable form, since

$$bb' - aa' = bb' - ab' + ab' - aa' = (b-a)b' + (b'-a')a.$$

DEFINITION. least residue

If $m > 0$ and r is the remainder when the division algorithm is used to divide b by m, then r is called the **least residue of** b **modulo** m.

Examples. The least residue of 12 modulo 7 is 5. The least residue of 7 modulo 12 is 7. The least residue of 20 modulo 4 is 0. The least residue of -12 modulo 7 is 2. The least residue of -3 modulo 7 is 4. ◇

Theorem 1.17. *Let* $m > 0$.

1. *If* r *and* b *are integers such that* $r \equiv b \pmod{m}$ *and* $0 \le r < m$, *then* r *is the least residue of* b *modulo* m.
2. *Two integers are congruent modulo* m *if and only if they have the same least residue modulo* m.

Proof.

1. Let $b - r = mq$. Then $b = mq + r$ with $0 \le r < m$ and r is the remainder in the division algorithm by the uniqueness of that remainder.

2. Suppose b and b' have the same remainder when divided by m, say

$$b = mq + r, \quad \text{and} \quad b' = mq' + r.$$

Then $b' - b = m(q' - q)$, so $b \equiv b' \pmod{m}$.

Conversely suppose $b \equiv b' \pmod{m}$ and $b = mq + r$, $0 \le r < m$. Then $b' \equiv r \pmod{m}$ by Theorem 1.16(3), so r is also the least residue of b' modulo m by part (1).

□

Example. Find the least residue (mod 31) of $33 \cdot 26^2$.

We have $33 \equiv 2 \pmod{31}$ and $26 \equiv -5 \pmod{31}$. By using part (5) of Theorem 1.16 twice we conclude that

$$33 \cdot 26^2 \equiv 2(-5)^2 = 50 \equiv 19 \pmod{31}.$$

Since $0 \le 19 < 31$, by Theorem 1.17 the least residue is 19. ◇

The congruence properties of Theorem 1.16 allow the substitution of congruent values in a polynomial with integer coefficients. Suppose the least residue of $3(53) + 27^2 \pmod{7}$ is desired. Since $53 \equiv 4 \pmod{7}$ and $27 \equiv -1 \pmod{7}$, we have

$$3(53) + 27^2 \equiv 3(4) + (-1)^2 = 13 \equiv 6 \pmod{7}, \tag{1.2}$$

and the least residue is 6. This justification for the first congruence of (1.2) is as follows, where the modulus is 7:

$$\begin{aligned}
53 &\equiv 4, \\
3(53) &\equiv 3(4) \quad (\text{part } (5)), \\
27 &\equiv -1, \\
27^2 = 27 \cdot 27 &\equiv (-1)(-1) = (-1)^2 \quad (\text{part } (5)), \\
3(53) + 27^2 &\equiv 3(4) + (-1)^2 \quad (\text{part } (4)).
\end{aligned}$$

Although both sides of a congruence may be multiplied by the same number by part (5) of Theorem 1.16, division is another matter. For example

$$2 \cdot 1 \equiv 2 \cdot 3 \pmod{4},$$

but $1 \not\equiv 3 \pmod{4}$. The following theorem tells when division is allowed.

Theorem 1.18 (the cancellation theorem). *If a, $b > 0$, x, and x' are integers such that $(a, b) = 1$, then $ax \equiv ax' \pmod{b}$ implies $x \equiv x' \pmod{b}$.*

This theorem is a special case of the following more general theorem.

Theorem 1.19. *If a, $b > 0$, d, x, and x' are integers such that $(a, b) = d$, then*

$$ax \equiv ax' \pmod{b} \quad \text{if and only if} \quad x \equiv x' \pmod{b/d}.$$

Proof. If $ax \equiv ax' \pmod{b}$, then b divides $a(x' - x)$. Say $a(x' - x) = bk$. Then $(a/d)(x' - x) = (b/d)k$. Since $(a/d, b/d) = 1$ by Theorem 1.14, b/d divides $x' - x$ by Theorem 1.13.

Conversely, if $x \equiv x' \pmod{b/d}$, then $x' - x = (b/d)k$ for some k. Then $ax' - ax = b(a/d)k$, which shows that $ax \equiv ax' \pmod{b}$. □

A Congruence Property of (a, b)

We return to our observation of the periodicity of (a, b) as b runs through the integers, made at the beginning of this section.

Theorem 1.20. *If $a > 0$, b, and b' are integers such that $b \equiv b' \pmod{a}$, then $(a, b) = (a, b')$.*

Proof. Let $b' - b = qa$. Then $b' = qa + b$. By Lemma 1.6, $(b', a) = (a, b)$. □

Problems for Section 1.5.

A

In the first eight problems, tell whether each statement is true or false.

1. $5 \equiv 27 \pmod{11}$
2. $63 \equiv 15 \pmod{9}$
3. $101 \equiv 29 \pmod{16}$
4. $-5 \equiv 43 \pmod{12}$
5. $7 \equiv -34 \pmod{9}$
6. $-50 \equiv 2 \pmod{13}$
7. $17 \equiv 62 \pmod{90}$
8. $-73 \equiv -29 \pmod{128}$

In the next 18 problems, find the least residue of b modulo m. Problems 17 through 22 are for students with calculators; see the method of Section 1.2. Theorem 1.16 may be useful in problems 23 through 26.

9. $m = 7$, $b = 100$
10. $m = 8$, $b = 77$
11. $m = 50$, $b = 17$
12. $m = 51$, $b = 19$
13. $m = 50$, $b = -12$
14. $m = 51$, $b = -30$
15. $m = 7$, $b = -100$
16. $m = 8$, $b = -77$
17. $m = 453$, $b = 613{,}571$
18. $m = 461$, $b = 916{,}215$
19. $m = 61{,}462$, $b = 818{,}886$
20. $m = 9162$, $b = 201{,}563$
21. $m = 41{,}522$, $b = -16{,}115$
22. $m = 91{,}631$, $b = -2152$
23. $m = 63$, $b = 752 \cdot 571$
24. $m = 51$, $b = 414 \cdot 566$

25. $m = 71$, $b = 72 \cdot 73 \cdot 74$

26. $m = 82$, $b = 80 \cdot 81 \cdot 85$

27. What is the least residue of 100^6 modulo 49?

28. What is the least residue of 49^4 modulo 23?

B

29. Suppose $a \equiv 2 \pmod{17}$, $b \equiv 4 \pmod{17}$, and $c \equiv 5 \pmod{17}$. What is the least residue of $a + bc \pmod{17}$?

30. With the assumptions of the previous problem, what is the least residue of $a^2 + b^2 + c^2 \pmod{17}$?

31. What is the least residue of $50^{99} \pmod{7}$?

32. What is the least residue of $50^{99} \pmod{17}$?

In the next three problems, use Theorem 1.19 to give a congruence equivalent to the one given with the numbers as small as possible.

33. $6x \equiv 14y \pmod{10}$

34. $6x \equiv 14y \pmod{15}$

35. $12x \equiv 30y \pmod{15}$

36. Let r and r' be the least residues of b and $-b$ modulo m. What is $r + r'$?

True-False. In the next five problems, tell whether each statement is true or false, and give counterexamples for the false statements. Write $a \sim b$ in case $(a, b) = 1$. Assume that a, b, c and d are nonzero integers.

37. Always $a \sim a$.

38. If $a \sim b$, then $b \sim a$.

39. If $a \sim b$ and $b \sim c$, then $a \sim c$.

40. If $a \sim b$ and $c \sim d$, then $a + c \sim b + d$.

41. If $a \sim b$ and $c \sim d$, then $ac \sim bd$.

42. Prove or disprove: If $c > 0$ and $a < b < c$, then $a \not\equiv b \pmod{c}$.

C

43. Give another proof of part (5) of Theorem 1.16 by first showing that if $a \equiv b \pmod{m}$, then $ac \equiv bc \pmod{m}$ for any c.

44. Show that if $x \equiv y \pmod{m}$, then

$$ax^2 + bx + c \equiv ay^2 + by + c \pmod{m}$$

for any integers a, b, and c.

45. Show that if $a \equiv a' \pmod{m}$ and $b \equiv b' \pmod{m}$, then $a - b \equiv a' - b' \pmod{m}$.

46. Prove that if m and n are positive integers and if $ab \equiv 1 \pmod{m}$ and $a^n \equiv 1 \pmod{m}$, then $b^n \equiv 1 \pmod{m}$.

47. Prove that if n is any integer, then $n \equiv 2 \pmod{3}$ or $n^2 \equiv n \pmod{6}$.

48. Let $m_1 = 2$, $m_2 = 3$, $m_3 = 4$, $m_4 = 6$, and $m_5 = 12$. Find integers a_1, a_2, a_3, a_4, a_5 such that if n is any integer, then there exists i such that $n \equiv a_i \pmod{m_i}$.

1.6 Mathematical Induction

Theorem 1.21. *If a, b, and $m > 0$ are integers such that $a \equiv b \pmod{m}$, then $a^n \equiv b^n \pmod{m}$ of all positive integers n.*

Proof. By part (5) of Theorem 1.16 we can multiply the congruence $a \equiv b \pmod{m}$ by itself, getting

$$a^2 \equiv b^2 \pmod{m}.$$

Multiplying this by the original congruence yields

$$a^3 \equiv b^3 \pmod{m}.$$

Clearly this process can be repeated as many times as desired, until we have

$$a^n \equiv b^n \pmod{m}.$$

□

Analysis of the Proof

The proof given above is convincing because it is obvious that one can always take the next step, from $a \equiv b$ to $a^2 \equiv b^2$ to $a^3 \equiv b^3$, and so on. Other statements true for all positive integers n, however, even though easy enough to confirm for any particular value of n, may not be so obviously true in general. Consider, for example, the following theorem.

Proposition A. *If n is any positive integer, then $4^n \equiv 3n + 1 \pmod{9}$.*

If $n = 1$, this says

$$4 \equiv 3 + 1 \pmod{9},$$

which is true. Likewise for $n = 2$ the congruence is

$$16 \equiv 6 + 1 \pmod{9},$$

and for $n = 3$ it is

$$64 \equiv 9 + 1 \pmod{9},$$

which are both also true. These particular cases do not convince us of the general truth of Proposition A, however. For this we need a more formal proof structure, that of mathematical induction.

Induction is a general method of proving facts to be true for each positive integer. The student usually is introduced to it as a means of proving formulas for sums, for example,

$$1 + 2 + \cdots + n = \frac{n(n+1)}{2}.$$

These sum formulas are often the only applications of induction the student sees, and he or she may form the opinion that its usefulness is quite limited. Nothing could be further from the truth. In this section we will give examples of induction proofs of various types. We start by formally stating the induction principle.

The principle of mathematical induction. Let $S(n)$ be a statement involving the positive integer n. Suppose

1. $S(1)$ is true.
2. If $S(k)$ is true for some integer k, then $S(k+1)$ is also true.

Then $S(n)$ is true for all positive integers n.

We illustrate the principle by using it to prove Theorem 1.21 and Proposition A.

Theorem 1.21. *If a, b, and $m > 0$ are integers such that $a \equiv b \pmod{m}$, then $a^n \equiv b^n \pmod{m}$ for all positive integers n.*

Proof. (Induction) Let $S(n)$ be the statement

$$a^n \equiv b^n \pmod{m}.$$

1. The statement $S(1)$ says

$$a \equiv b \pmod{m}, \text{ which is true by hypothesis.}$$

2. Suppose $S(k)$ is true, that is

$$a^k \equiv b^k \pmod{m}.$$

Multiplying this by $a \equiv b \pmod{m}$ yields $a^{k+1} \equiv b^{k+1} \pmod{m}$, which is $S(k+1)$.

We have verified (1) and (2) of the induction principle. Then by the principle $a^n \equiv b^n \pmod{m}$ for all positive integers n. □

Proposition A. *If n is any positive integer, then $4^n \equiv 3n+1 \pmod{9}$.*

Proof. (Induction) Let $S(n)$ be the statement

$$4^n \equiv 3n+1 \pmod{9}.$$

1. The statement $S(1)$ is

$$4 \equiv 3+1 \pmod{9}, \text{ which is true.}$$

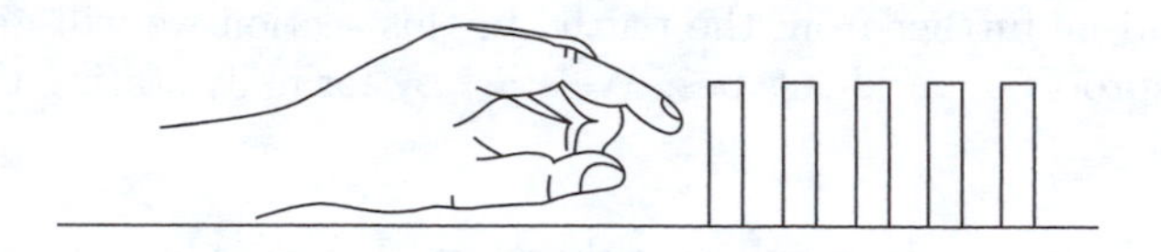

Figure 1.1

2. Assume $S(k)$, that is
$$4^k \equiv 3k + 1 \pmod{9}.$$
Clearly $4 \equiv 4 \pmod{9}$. Multiplying these congruences gives
$$4^{k+1} \equiv 4(3k + 1) \pmod{9}.$$
The right side of this congruence is
$$(3 + 1)(3k + 1) = 9k + 3(k + 1) + 1.$$
Thus
$$4(3k + 1) \equiv 3(k + 1) + 1 \pmod{9},$$
since subtracting the sides of this congruence gives $-9k$. By Theorem 1.16 part (3) we have
$$4^{k+1} \equiv 3(k + 1) + 1 \pmod{9},$$
and this is $S(k + 1)$.

We have verified (1) and (2), and so by the principle of mathematical induction $4^n \equiv 3n + 1 \pmod{9}$ for all positive integers n. □

Is Induction True?

Students often have trouble accepting the induction principle. The usual objection is that in part (2) we are "assuming what we are trying to prove." This is not the case, and the confusion arises because of inattention to the domain of the variable n. We are trying to prove that $S(n)$ holds for all positive integers n. In part (2) we assume merely that $S(k)$ holds for one particular k, and show that $S(k+1)$ also holds. This assumption is not unreasonable, since we already know it to be correct for $k = 1$.

The induction principle is analogous to a string of dominoes, set up in a single line (see Figure 1.1). Suppose that

1. I knock down the first domino.
2. The dominoes are so arranged that if one falls down, then so does the next one.

Then every domino will fall down, no matter how many there are.

More Induction Proofs

We will now give more examples of induction proofs. As is customary, the statement $S(n)$ will not be labeled explicitly as such.

Proposition B. *If X is any real number greater than -1, then*

$$(1+X)^n \geq 1+nX$$

for all positive integers n.

Proof. (Induction)

1. If $n=1$ the inequality is

$$1+X \geq 1+X, \text{ which is clearly true.}$$

2. Suppose $(1+X)^k \geq 1+kX$. Since $1+X>0$ we can multiply through by $1+X$ to get

$$\begin{aligned}(1+X)^{k+1} &\geq (1+X)(1+kX)\\ &= 1+(k+1)X+kX^2 \geq 1+(k+1)X.\end{aligned}$$

This is the inequality of $n=k+1$.

Thus by the induction principle $(1+X)^n \geq 1+nX$ for all positive integers n. □

Theorem 1.22. *If the integers $a_1, a_2, \ldots, a_n$ are all relatively prime to b, then so is their product $a_1a_2\cdots a_n$.*

Proof. The proof will be by induction on n.

1. If $n=1$, then we are given that $(a_1, b)=1$, which is also what is to be proved.

2. Now assume the theorem is true for some integer k, and that $a_1, a_2, \ldots, a_k$, and a_{k+1} are all relatively prime to b. By the induction hypothesis $(a_1a_2\cdots a_k, b)=1$, and it is given that $(a_{k+1}, b)=1$. Then by Corollary 1.12 we have $(a_1\cdots a_ka_{k+1}, b)=1$.

Thus by the induction principle the theorem hold for all positive integers n. □

Theorem 1.23 (the sum of a geometric progression). *For any real numbers A and $R \neq 1$ and positive integer n*

$$A+AR+AR^2+\cdots+AR^{n-1}=\frac{A(R^n-1)}{R-1}.$$

Proof. (Induction)

1. If $n = 1$, the formula says $A = A(R-1)/(R-1)$, which is true.

2. Assume that for some positive integer k we have

$$A + AR + AR^2 + \cdots + AR^{k-1} = \frac{A(R^k - 1)}{R - 1}.$$

Adding AR^k to both sides gives

$$A + AR + AR^2 + \cdots + AR^{k-1} + AR^k = \frac{A(R^n - 1)}{R - 1} + AR^k$$

$$\begin{aligned} &= A\left(\frac{R^k - 1}{R - 1} + \frac{R^k(R-1)}{R-1}\right) = A\left(\frac{R^k - 1 + R^{k+1} - R^k}{R - 1}\right) \\ &= A\left(\frac{R^{k+1} - 1}{R - 1}\right), \end{aligned}$$

which is the formula for $n = k + 1$.

Thus by the principle of mathematical induction, the formula for the sum of a geometric progression holds for all positive integers n. □

DEFINITION. Fibonacci number

We define a sequence of number as follows:

$$\begin{aligned} f_1 &= 1, \\ f_2 &= 1, \\ f_3 &= 1 + 1 = 2, \\ f_4 &= 1 + 2 = 3, \\ f_5 &= 2 + 3 = 5, \end{aligned}$$

and, in general, if $n \geq 3$, then

$$f_n = f_{n-2} + f_{n-1}.$$

The number f_n is called the nth **Fibonacci number**.

The following table gives the first 16 Fibonacci numbers.

n	1	2	3	4	5	6	7	8	9	10	11	12	13	14	15	16
f_n	1	1	2	3	5	8	13	21	34	55	89	144	233	377	610	987

Various relationships among these numbers my be discovered by trial and error and proved by induction. Let us, for example, add up initial groups of Fibonacci numbers with even subscripts.

$f_2 = 1$
$f_2 + f_4 = 1 + 3 = 4$
$f_2 + f_4 + f_6 = 1 + 3 + 8 = 12$
$f_2 + f_4 + f_6 + f_8 = 1 + 3 + 8 + 21 = 33$
$f_2 + f_4 + f_6 + f_8 + f_{10} = 1 + 3 + 8 + 21 + 55 = 88$

The reader may notice that 1, 4, 12, 33, and 88 are 1 less than 2, 5, 13, 34, and 89, all Fibonacci numbers with odd subscripts.

Proposition C. *For each positive integer n*

$$f_2 + f_4 + \cdots + f_{2n} = f_{2n+1} - 1.$$

Proof. (Induction)

1. If $n = 1$ the formula says $f_2 = f_3 - 1$, or $1 = 2 - 1$.
2. Assume $f_2 + f_4 + \cdots + f_{2k} = f_{2k+1} - 1$ for some positive integer k. Adding $f_{2(k+1)}$ to both sides gives

$$f_2 + f_4 + \cdots + f_{2k+2} = f_{2k+1} + f_{2k+2} - 1 = f_{2k+3} - 1,$$

where the last equation is by the definition of the Fibonacci sequence. Since this is $f_{2(k+1)+1} - 1$, we have proved the formula for $n = k + 1$.

Thus by induction the formula is true for all positive integers n. □

Problems for Section 1.6.

A

In the first 14 problems problems, tell whether or not each sequence given is a geometric progression. If it is, tell what A, R, and n are, and use Theorem 1.23 to compute its sum.

1. $2, 4, 8, 16, 32$
2. $3, 1, \frac{1}{3}, \frac{1}{9}$
3. $-\frac{1}{2}, \frac{1}{2}, \frac{3}{2}, \frac{5}{2}$
4. $3, -3, 3, -3, 3, -3$
5. $-1, 1\frac{1}{2}, 4$
6. $1, \frac{1}{2}, \frac{1}{3}, \frac{1}{4}, \frac{1}{5}$
7. $-\frac{1}{2}, 1, -2$
8. $1, -2, 3, -4, 5$
9. $1, 4, 9, 16, 25$
10. $0.1, 0.01, 0.001$
11. $1, \frac{1}{3}, \frac{1}{9}, \frac{1}{27}, \frac{1}{81}, \frac{1}{243}$
12. $-1, 2, -4, 8, -16, \ldots, 512$
13. $\frac{3}{2}, \frac{2}{3}, \frac{8}{27}, \frac{32}{243}$
14. $3, 2.5, 2, \ldots, -7.5$
15. Compute f_n for $n = 17$ through 20.
16. Compute $f_1 + f_2 + \cdots + f_n$ for $n = 1, 2, 3, 4, 5, 6$. Conjecture a simpler formula for the sum.

B

In problems 17 through 53, the statement is to be proved true for all positive integers n by induction.

17. $1 + 4 + 9 + \cdots + n^2 = n(n+1)(2n+1)/6$.
18. $1 + 8 + 27 + \cdots + n^3 = n^2(n+1)^2/4$.

19. $7^n \equiv 6n + 1 \pmod{36}$.
20. $2 \cdot 7^n \equiv 2^n(2 + 5n) \pmod{25}$.
21. $2^n + 3^n \equiv 5^n \pmod{6}$.
22. $16^n \equiv 1 - 10n \pmod{25}$.
23. $1 \cdot 1! + 2 \cdot 2! + \cdots + n \cdot n! = (n+1)! - 1$.
24. $3 \mid 4^n - 1$.
25. $1/(1 \cdot 2) + 1/(2 \cdot 3) + \cdots + 1/n(n+1) = n/(n+1)$.
26. $1/(1 \cdot 3) + 1/(3 \cdot 5) + \cdots + 1/(2n-1)(2n+1) = n/(2n+1)$.
27. $2^n \mid (n+1)(n+2) \cdots (2n)$.
28. If $a \mid b_i$ for $i = 1, \ldots, n$, and if $x_1, x_2, \ldots, x_n$ are integers, then $a \mid x_1b_1 + x_2b_2 + \cdots + x_nb_n$.
29. If $a \mid b_1b_2 \cdots b_nc$, and if $(a, b_i) = 1$ for $i = 1, \ldots, n$, then $a \mid c$.
30. If $a_i \equiv b_i \pmod{m}$ for $i = 1, \ldots, n$, then $a_1 + a_2 + \cdots + a_n \equiv b_1 + b_2 + \cdots + b_n \pmod{m}$.
31. $f_1 + f_3 + f_5 + \cdots + f_{2n-1} = f_{2n}$.
32. $f_1 + f_2 + \cdots + f_n = f_{n+2} - 1$.
33. $f_n < 2^n$.
34. If A is the matrix $\begin{bmatrix} 1 & 1 \\ 1 & 0 \end{bmatrix}$, then $A^{n+1} = \begin{bmatrix} f_{n+2} & f_{n+1} \\ f_{n+1} & f_n \end{bmatrix}$.
35. $2^{2^n} + 1 \equiv 5 \pmod{12}$.
36. A set with n elements has exactly 2^n subsets.
37. (For those familiar with calculus) $(d/dx)(x^n) = nx^{n-1}$. Assume only $(d/dx)(x) = 1$ and the product rule for differentiation.
38. $(-4)^n \equiv 1 - 5n \pmod{25}$.
39. $2^{3^n} \equiv -1 \pmod{3^n}$
40. $(\frac{1}{2} + 1)(\frac{1}{2} + \frac{1}{2})(\frac{1}{2} + \frac{1}{3}) \cdots (\frac{1}{2} + \frac{1}{n}) = (n+1)(n+2)/2^{n+1}$.
41. $2^{2n-1}(n!)^2 \geq (2n)!$.
42. $5^n \equiv 1 + 4n \pmod{16}$.
43. $5^n \equiv 8n^2 - 4n + 1 \pmod{64}$.
44. $8^n \mid (4n)!$.
45. $(2n)! \geq (n!)^2 2^n$.
46. $2^{n+2} \mid (2n+3)!$.
47. $3^{2n+1} \equiv 2(-1)^{n+1} \pmod{5}$.
48. $16^n \mid (6n)!$.
49. $(2n)!/(n!)^2 \leq 4^n$.
50. $(1!1^1)(2!2^2) \cdots (n!n^n) = (n!)^{n+1}$.
51. $-1 + 2 - 3 + \cdots + (-1)^n n = ((-1)^n(2n+1) - 1)/4$.
52. $31 \mid 2^{5n} - 1$.
53. $17 \mid 2^{8n-4} + 1$.

54. Prove that $2^n \mid n(n+1) \cdots (2n)$ for all positive integers n. (This is not the same as problem 27.)

C

55. Let n_0 be a fixed integer, and let $S(n)$ be a statement involving n. Suppose that (1) $S(n_0)$ is true and (2) whenever $S(k)$ is true for $k \geq n_0$, then $S(k+1)$ is true. Prove that $S(n)$ is true for all $n \geq n_0$.
56. Prove that $(2n)! < (n!)^2 4^{n-1}$ for $n \geq 5$.
57. The **well-ordering principle** states that each nonempty set of positive integers contains a least element. Assume the well-ordering principle and prove from it the principle of mathematical induction. (**Hint:** Suppose (1) and (2) of the induction principle hold. Let n be the least positive integer such that $S(n)$ is false.)
58. Give a formal induction proof of the Euclidean algorithm.

59. What is wrong with the following proof that all things are the same? We prove by induction on n that any n things are the same. If $n = 1$ this is clear. Assume any k things are the same. Then given $k + 1$ things $a_1, a_2, \ldots, a_{k+1}$, we have by the induction hypothesis that $a_1 = a_2 = \cdots = a_k$ and $a_2 = a_3 = \cdots = a_{k+1}$. Thus $a_1 = a_2 = \cdots a_{k+1}$. Thus by the principle of mathematical induction all things are the same.
60. Prove that if a, m, and n are positive integers, and n is odd, then $(a^n - 1, a^m + 1) = 1$ or 2. (**Hint:** Theorem 1.21.)
61. Show that if $n^2 = a^2 + b^2 + c^2$, and n is even, then a, b, and c are all even. (**Hint:** Show that $x^2 \equiv 0$ or $1 \pmod 4$ for all x.)
62. Prove by induction on n that if n is any positive integer then $4^n = a^2 + b^2 + c^2$ is impossible for positive integers a, b, and c.
63. Let a_1, a_2, ... be a sequence of integers such that $a_1 = 1$ and $a_n = 2a_{n-1} + 1$ for $n \geq 2$. Prove that $a_n = 2^n - 1$ for all positive integers n.
64. If S is a set, we denote the number of elements it has by $|S|$. Prove that if $|S| = n$, then $\sum_{T \subseteq S} 2^{|T|} = 3^n$.

Pierre de Fermat

1608–1665

The great French scientific amateur Pierre de Fermat led a quiet life practicing law in Toulouse, and producing high-quality work in number theory and other areas of mathematics as a hobby. He published almost nothing, revealing most of his results in his extensive correspondence with friends, and generally kept his proofs to himself. "Fermat's theorem" (not the same as Fermat's *last* theorem), which will be presented in Section 4.3, first appeared in a letter to Frenicle de Bessy dated October 18, 1640.

Fermat also worked with differential calculus, finding a way of calculating maxima and minima. He reasoned that if f had an extremum at x, then changing x slightly would not have much effect on $f(x)$. For example, if $f(x) = x^2 + x$, then, letting e represent the change in x, he set $f(x+e) = f(x)$. This gives

$$x^2 + 2xe + e^2 + x + e = x^2 + x,$$

or, after dividing by e,

$$2x + e + 1 = 0.$$

Now setting $e = 0$ gives $x = -\frac{1}{2}$, which is the x minimizing $f(x)$.

Calculus students will recognize that this amounts to calculating the derivative of f from the definition and setting it equal to 0. The reason most people do not credit Fermat with being the founder of differential calculus (as was claimed by his countrymen Lagrange, Laplace, and Fourier) is that, although Fermat could solve individual problems such as the one given, he did not develop general differentiation rules as did Newton and Leibniz. Nevertheless, in a letter that came to light only in 1934, Newton did acknowledge that he got the hint that led to the development of the differential calculus from Fermat's method of constructing tangents.

PIERRE DE FERMAT

Chapter 2

Prime Numbers

2.1 Prime Factorization

DEFINITION. prime, composite

A positive integer is said to be **prime** in case it has exactly two positive divisors. A positive integer having more than two positive divisors is said to be **composite**. The number 1 is neither prime nor composite.

Examples. The primes less than 20 are 2, 3, 5, 7, 11, 13, 17, and 19. ◇

Theorem 2.1. *Suppose $n > 1$. Then n can be written as the product of primes.*

Proof. If n is prime we are done. (We consider a prime to be a product of one prime.) Otherwise n must have a positive divisor, say d, other than itself and 1. Then $1 < d < n$. Let $n = dd'$. Clearly $1 < d' < n$ also. We now apply the same argument to d and d' as we did to n. This procedure must end, since the factors grow smaller at each step. But it can stop only when each factor has no positive divisors other than 1 and itself, that is, when each factor is prime. □

Illustration of the proof. Suppose $n = 120$. We know that $120 = 12 \cdot 10$. Now $12 = 3 \cdot 4$ and $4 = 2 \cdot 2$. Thus $12 = 3 \cdot 2 \cdot 2$, and all these factors are prime. Likewise $10 = 2 \cdot 5$, and 2 and 5 are prime. Finally,

$$120 = 3 \cdot 2 \cdot 2 \cdot 2 \cdot 5,$$

and 3, 2, and 5 are prime.

Comment on the proof. The proof given is convincing because we can see how each step will proceed, even though the number of steps and branches involved will depend on the particular value of n. In this respect, it is like our first proof of Theorem 1.21 (showing that $a \equiv b$ implies $a^n \equiv b^n$). We suspect that a more formal argument using mathematical induction could be given. This is the case; but a slightly different form of the induction principle must be used.

The principle of mathematical induction II. Let $S(n)$ be a statement involving the positive integer n. Suppose

1. $S(1)$ is true.
2. If for some positive integer k all of $S(1), S(2), \ldots, S(k)$ are true, then $S(k+1)$ is true.

Then $S(n)$ is true for all positive integers n.

Comment on induction II. Of course, the difference between induction II and the form of induction of Section 1.6 is entirely in condition (2). With induction II we are allowed to assume not only that $S(k)$ is true, but also $S(1), S(2), \ldots, S(k-1)$. Since we are allowed to assume more, an induction II proof is actually easier, at least from a logical standpoint. Induction I (which what we will call the principle of Section 1.6 when we want to distinguish between the two forms) is, however, formally simpler and suffices for most proofs.

The reader should study the induction proof we now give for Theorem 2.1 and decide why induction II is needed in it.

Theorem 2.1. *Suppose $n > 1$ is an integer. Then n can be written as a product of primes.*

Proof. The proof will be by induction II.

1. If $n = 1$ there is nothing to prove, since the theorem concerns only values of $n > 1$.

2. Suppose we know the theorem is true for $n = 1, 2, \ldots, k$. Consider the integer $k+1$. If $k+1$ is prime, then we are done. Otherwise

$$k + 1 = d \cdot d',$$

where $1 < d < k+1$ and $1 < d' < k+1$. Since d and d' are each at most k, by the induction assumption each can be written as a product of primes. Thus n can also be written as such a product.

Thus by induction II any $n > 1$ can be written as a product of primes. □

We will give two more examples of induction II proofs. Recall that f_n is the nth Fibonacci number.

Theorem 2.2. *Let $A = (\sqrt{5}+1)/2$. Then $f_{n+2} > A^n$ for all positive integers n.*

Proof. The proof will be by induction II. Note that A is a solution to the equation $x^2 - x - 1 = 0$, and so $A + 1 = A^2$.

1. If $n = 1$, we have

$$f_{n+2} = f_3 = 2 = \frac{3+1}{2} > \frac{\sqrt{5}+1}{2} = A^n,$$

and so the inequality holds.

We will also check the case $n = 2$ so that we may assume that $k > 1$ in the second part of this proof. Then

$$f_{n+2} = f_4 = 3 = \frac{3+3}{2} > \frac{\sqrt{5}+3}{2} = A + 1 = A^2 = A^n.$$

2. Suppose we know for some integer $k > 1$ the inequality holds for $n = 1, 2, \ldots, k$. Then

$$f_{(k+1)+2} = f_{k+2} + f_{k+1} > A^k + A^{k-1} = A^{k-1}(A+1) = A^{k-1}A^2 = A^{k+1}.$$

Thus by Induction II the theorem holds for all positive integers n. □

The next result was mentioned in Section 1.5. It justifies the "substitution" of congruent numbers in integral polynomials.

Theorem 2.3. *Let $P(x)$ be a polynomial with integer coefficients, and let u, v, and $m > 0$ be integers such that $u \equiv v \pmod{m}$. Then $P(u) \equiv P(v) \pmod{m}$.*

Proof. If $P(x)$ is constant there is nothing to prove, so we will assume $P(x)$ has degree $n \geq 1$. Suppose $P(x) = a_n x^n + a_{n-1}x^{n-1} + \cdots + a_0$, where $a_0, a_1, \ldots, a_n$ are integers with $a_n \neq 0$. The proof will be by induction II on n.

1. If $n = 1$, then $P(x) = a_1 x + a_0$. Then by parts (1), (5), and (4) of Theorem 1.16

$$\begin{aligned} a_0 &\equiv a_0 \pmod{m} \\ a_1 &\equiv a_1 \pmod{m} \\ a_1 u &\equiv a_1 v \pmod{m} \\ P(u) = a_1 u + a_0 &\equiv a_1 v + a_0 = P(v) \pmod{m}. \end{aligned}$$

2. Now assume the statement of the theorem is true for all polynomials with degrees $\leq k$. Suppose $n = k + 1$, so $P(x) = a_{k+1}x^{k+1} + a_k x^k + \cdots + a_0$. Let $Q(x) = a_k x^k + \cdots + a_0$. Now $Q(x)$ is either constant or has degree $\leq k$, so

$$Q(u) \equiv Q(v) \pmod{m}.$$

Also $u^{k+1} \equiv v^{k+1} \pmod{m}$ by Theorem 1.21, so $a_{k+1}u^{k+1} \equiv a_{k+1}v^{k+1} \pmod{m}$ as above. Thus

$$P(u) = a_{k+1}u^{k+1} + Q(u) \equiv a_{k+1}v^{k+1} + Q(v) = P(v) \pmod{m}.$$

This proves our result for $n = k + 1$.

Thus by induction II our theorem holds for integral polynomials of all positive degrees, and so for all integral polynomials. □

The Number of Divisors of an Integer

Theorem 2.1 is very important, but gives only half of the story of factorization into primes. To motivate the other half we consider counting the positive divisors of an integer. It will turn out that its factorization into primes will tell us how to do this.

DEFINITION. $\tau(n)$

We define $\tau(n)$ to be the number of positive divisors of n.

Examples.

n	positive divisors of n	$\tau(n)$
1	1	1
2	1, 2	2
3	1, 3	2
4	1, 2, 4	3
5	1, 5	2
6	1, 2, 3, 6	4
7	1, 7	2
8	1, 2, 4, 8	4
9	1, 3, 9	3
10	1, 2, 5, 10	4

◇

We will examine the function $\tau(n)$ in more detail in the next section.

Problems for Section 2.1.

1. What are the primes between 20 and 40?
2. What are the primes between 40 and 60?
3. What are the primes between 100 and 120?
4. What are the primes between 120 and 140?
5. Compute $\tau(n)$ for n between 11 and 20.
6. Compute $\tau(n)$ for n between 21 and 30.

In the next four problems, write n as a product of primes. Write the primes in nondecreasing order. For example, $120 = 2 \cdot 2 \cdot 2 \cdot 3 \cdot 5$.

7. $n = 180$ 8. $n = 500$ 9. $n = 1001$ 10. $n = 4199$

In the next four problems, compute $\tau(n)$ for each given n.

11. $n = 75$ 12. $n = 99$ 13. $n = 243$ 14. $n = 1024$

15. Compute f_3 using the formula of problem 32.
16. Compute f_4 using the formula of problem 32.
17. What do you think $\tau(pq^2)$ is, given that p and q are distinct primes?
18. What do you think $\tau(p^2q^3)$ is, given that p and q are distinct primes?
19. What is $\tau(2^n)$?
20. What is $\tau(2^n 3^m)$?

21. Find the smallest $n > 0$ such that $\tau(n) = 5$.
22. Find the smallest $n > 0$ such that $\tau(n) = 6$.
23. Find the smallest n such that $\tau(n) = 100$.
24. Find an n not a power of a prime such that $\tau(n) = 51$.

C

In the next eight problems, prove the statement given is true for all positive integers n.

25. $f_1^2 + f_2^2 + \ldots + f_n^2 = f_n f_{n+1}$.
26. $f_n + f_{n+1} + f_{n+3} = f_{n+4}$.
27. Let s_n be the number of subsets of $\{1, 2, \ldots, n\}$ not containing consecutive integers. For example, $s_1 = 2$, $s_2 = 3$, and $s_5 = 13$. Then $s_n = f_{n+2}$.
28. $f_1 - f_2 + f_3 - \ldots + (-1)^n f_{n+1} = (-1)^n f_n + 1$.
29. If $0 < k < n$, then $f_{n+1} = f_{n-k} f_k + f_{n-k+1} f_{k+1}$. (**Hint:** Fix k and use induction II on n starting with $n = k + 1$.)
30. $f_n < (5/3)^n$.
31. $2^n f_n < (\sqrt{5} + 1)^n$.
32. $f_n = (A^n - B^n)/\sqrt{5}$, where A and B are the positive and negative solutions to $x^2 - x - 1 = 0$, respectively.

33. Let $a_1 = 1$, $a_2 = 3$, and $a_n = a_{n-1} + a_{n-2}$ for $n = 3, 4, \ldots$. Show that $a_n < (7/4)^n$ for all positive integers n.
34. Prove that if p is prime, then $\tau(p^n) > n$.
35. Suppose that p and q are distinct primes, and that p and q are the only primes dividing pq. Show that $\tau(pq) = 4$.
36. Show that induction II implies induction I.
37. Show that induction I implies induction II. (**Hint:** Suppose $S(n)$ is a statement satisfying (1) and (2) of induction II. Let $S'(n)$ be the statement "$S(k)$ is true for $k = 1, 2, \ldots, n$." Apply induction I to $S'(n)$.)
38. Show that induction II implies the well-ordering principle. (See the problems at the end of Section 1.6.) (**Hint:** Let T be a nonempty set of positive integers with no smallest element. Let $S(n)$ be the statement that n is not in T. Show that T is empty.)
39. Show that the sum of two consecutive primes (such as 19 and 23) is never twice a prime.
40. Prove Theorem 2.3 with induction I by using $a_{k+1}x^{k+1} + \cdots + a_0 = x(a_{k+1}x^k + \cdots + a_1) + a_0$.
41. Prove that if $\tau(n) = n$, then n is 1 or 2.

2.2 The Fundamental Theorem of Arithmetic

What Is $\tau(pq)$?

Recall that $\tau(n)$ denotes the number of positive divisors of n. It is easy enough for us to compute that $\tau(2 \cdot 3) = 4$, since the positive divisors of 6 are 1, 2, 3, and 6. In general, if p and q are distinct primes then pq has the divisors 1, p, q, and pq; and so $\tau(pq)$ is at least 4. The reader may feel it is obvious that $\tau(pq)$ is exactly 4. After all, what other divisors could pq have?

This is a situation where our familiarity with the integers may be a disadvantage, since it may lead us to assume as true things that really should be proved. It happens to be true that pq has no factors other than those listed, but we cannot let intuition and experience substitute for a rigorous proof of this fact. As useful as intuition is in mathematics, there are many instances where it has led to serious errors.

In order to demonstrate that one should not put too much trust in one's intuition about the integers, we will now present two examples of systems in which certain laws with which we are familiar do not hold.

The System E

Let E be the set of all even positive integers: 2, 4, 6, 8, ... If a and b are in E, we say that a **divides** b **in** E in case $b = ak$ for some element k in E. Of course, this parallels the usual definition of divisibility in the integers. For example, 2 does not divide 6 in E because although $6 = 2 \cdot 3$, 3 is not in E.

Since 1 is not in E we will have to revise our definition of primality; but the definition we will give will be equivalent to the previous one in the case of the ordinary integers. We will say that p in E is **prime** if p cannot be written as the product of two other elements of E greater than 1; for example, 2 and 6 are prime in E, but $4 = 2 \cdot 2$ is not.

The reader may easily check that 10 and 30 are also prime in E. Thus $60 = 6 \cdot 10$ is of the form pq, where p and q are distinct primes.

Now comes the shocker.

The integer $60 = 2 \cdot 30$, and 2 and 30 are primes. We have factored 60 into prime factors in two essentially different ways, something (it will turn out) we could never do with the ordinary integers.

Note that the system E is not at all exotic, and in fact shares most of the usual algebraic properties of the integers. For example, it is closed under both addition and multiplication. One basic property it lacks is that of having a multiplicative unit 1. Our next example will contain 1.

The System T

Let T be all positive integers congruent to 1 modulo 3; that is, the numbers 1, 4, 7, 10, ... Since $a \equiv 1 \pmod 3$ and $b \equiv 1 \pmod 3$ imply $ab \equiv 1 \pmod 3$ we see that T is closed under multiplication.

If a and b are in T we say a **divides** b **in** T in case $b = ak$ for some k in T, and call an element $p > 1$ **prime** if it cannot be written as the product of two factors in T greater than 1.

We find the smallest composite element of T to be $16 = 4 \cdot 4$, and it is easy to check that 4, 10, 22, and 55 are all prime elements of T.

Another shocker.

These numbers provide an example of nonunique prime factorization in T, since $220 = 4 \cdot 55 = 10 \cdot 22$. In fact 220 has the *six* divisors 1, 4, 10, 22, 55, and 220 in T, in spite of the fact that 220 is of the form pq for distinct primes p and q.

The Unique Factorization Property

Chastened by the above examples we return to the ordinary integers to prove a theorem showing that such aberrations do not occur among them.

Theorem 2.4. *Suppose a and b are integers, p is prime, and $p \mid ab$. Then $p \mid a$ or $p \mid b$.*

Proof. Suppose $p \nmid a$. Then, since the only positive divisors of p are p and 1, (a, p) must be 1. Thus $p \mid b$ by Theorem 1.13. □

This result is the key to showing that the ordinary integers enjoy unique factorization. Suppose the integer $n > 1$ has two factorizations into primes, say

$$n = p_1 p_2 \cdots p_s = q_1 q_2 \cdots q_t.$$

Since p_1 divides the right side of the last equation, Theorem 2.4 implies that p_1 divides either q_1 or $q_2 \cdots q_t$. In the latter case, p_1 divides either q_2 or the product of the remaining qs by the same argument. Eventually we find that p_1 divides one of the qs, say q_i. But since q_i is prime, its only divisors are itself and 1, so we must have $p_1 = q_i$.

Now we divide both sides of the equation by $p_1 = q_i$ and apply the same argument to p_2, finding that it must equal one of the remaining qs. We cancel this prime in the same way. This process is continued as long as possible. Both sides must be exhausted at the same time, otherwise we would have a product of primes equaling 1. We have matched each p with a q. Thus the qs must simply be a rearrangement of the ps.

Notice that the key to proving unique factorization is Theorem 2.4. This result may be found in Book VII of Euclid's *Elements*, stated as follows: "If two numbers by multiplying one another make some number, and any prime number measure the product, it will also measure one of the original numbers."

The theorem just proved is often combined with the theorem giving the existence of a prime factorization, as follows:

Theorem 2.5 (the fundamental theorem of arithmetic). *Any integer n greater than 1 has a factorization into primes. This factorization is unique up to the order of the factors.*

Problems for Section 2.2.

1. List the primes in E between 1 and 20.
2. List the primes in E between 21 and 40.
3. List the primes in T between 1 and 30.
4. List the primes in T between 31 and 60.

In the next four problems, find the number of divisors of n in E.

5. $n = 40$ 6. $n = 60$ 7. $n = 90$ 8. $n = 84$

In the next four problems, find the number of divisors of n in T.

9. $n = 64$ 10. $n = 85$ 11. $n = 250$ 12. $n = 550$

13. How many divisors does a prime in E have?
14. How many divisors does a prime in T have?
15. Find a number in T less than 220 that can be factored into primes in two essentially different ways.
16. Show that T is not closed under addition.
17. Give an example of distinct primes p, q, and r in E such that $p^2 = qr$.
18. Give an example of distinct primes p, q, and r in T such that $p^2 = qr$.

19. The set T consists of all positive integers of the form $3k + 1$. Use this to prove without using congruences that T is closed under multiplication.
20. Prove that E is closed under addition and multiplication.
21. Use Theorem 2.4 to show that if p and q are distinct primes, then $\tau(pq) = 4$.
22. Show by induction that if a prime p divides a product of n numbers, then it divides at least one of the numbers.
23. Prove that if n is in E, then n can be written as a product of primes in E. Use induction II.
24. Show that if the square of no prime divides b and $a^2 \mid bc^2$, then $a \mid c$.
25. Show that $a \mid b$ if and only if whenever p is prime and $p^k \mid a$, then $p^k \mid b$.
26. Use the previous problem to show that if a and b are positive integers, then $(a, b) = (a + b, [a, b])$.
27. Prove that if $n > 1$, then $n = mk^2$, where the square of no prime divides m.

2.3 The Importance of Unique Factorization

The assumption that the factorization of a number into primes must be unique has a famous precedent in mathematical history. Consider the equation

$$x^n + y^n = z^n, \quad x > 0,\ y > 0,\ z > 0. \tag{2.1}$$

This is an example of a **Diophantine equation**, that is, an equation for which integral (or, sometimes, rational) solutions are desired. The name comes from the Greek mathematician Diophantus, who studied many such equations.

If $n = 2$, it is easy to find solutions to (2.1); for example, $x = 3$, $y = 4$, $z = 5$, or $x = 12$, $y = 5$, $z = 13$. For $n > 2$, however, it is a different story. An account of the recent proof by Andrew Wiles that no solution exists if $n > 2$ may be found in Chapter 0 of this book. The proof settles a problem that goes back more than 350 years.

Around 1637 the French mathematician Pierre de Fermat, writing in the margin of his copy of the works of Diophantus, claimed that he had found a "truly wonderful" proof that (2.1) had no solution for $n > 2$, but that there was not room for him to write it down there. This statement is known as **Fermat's last theorem**. If Fermat wrote down his alleged proof anywhere, no one ever found it.

More than 200 years after Fermat's claim, interest developed in the connection between Fermat's last theorem and algebraic integers. These are real or complex numbers that behave like the ordinary integers in many respects. In particular, the concepts of divisibility and primality can be defined in various sets of algebraic integers. The French mathematicians Lamé and Cauchy thought they were close to proofs of Fermat's last theorem, based on the assumption that prime factorization of algebraic integers was unique.

An often-repeated story is that the German mathematician Kummer submitted a manuscript purporting to contain a proof, which was invalid because of this assumption. (Inside many classes of algebraic integers factorization is *not* unique; the set S that we will define shortly is one example.) Kummer's supposed manuscript has never been found, however, and doubt has been thrown on this part of the story. The interested reader should consult the book by Edwards in the references.

In any case, Kummer went on to prove many important results related to Fermat's last theorem, including the impossibility of (2.1) for a large number of values of n. His work led the way to the modern theory of algebraic numbers.

The System S

We define S to be the set of all complex numbers of the form

$$a + b\sqrt{-6},$$

where a and b are ordinary integers. It is easy to see that S is closed under addition and, since

$$(a + b\sqrt{-6})(c + d\sqrt{-6}) = ac - 6bd + (ad + bc)\sqrt{-6},$$

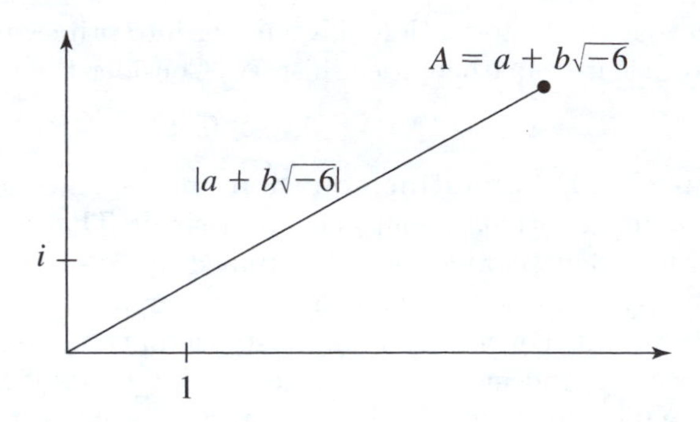

Figure 2.1 An element of S

also closed under multiplication.

As usual, if A and B are in S we say A **divides** B **in** S in case $B = AC$ for some element C of S. Since the concepts of being positive or negative do not apply to complex numbers, we must modify our definition of primality. We say P in S is **prime** in case we cannot write P as a product AB where A and B are in S and neither A nor B is 1 or -1.

It is perhaps not clear that elements of S can be factored into primes, since the argument given in Section 2.1 depended on the factors of a number being smaller than the number; while here the factors are complex numbers. To get around this, we consider the square of the modulus (distance from the origin) of an element $A = a + b\sqrt{-6}$ or S, given by

$$|a + b\sqrt{-6}|^2 = |a + b\sqrt{6}i|^2 = a^2 + 6b^2.$$

(See Figure 2.1.) It is not hard to check that

1. $|A|^2 = 0$ if and only if $A = 0$;
2. $|A|^2 = 1$ if and only if $A = 1$ or $A = -1$;
3. $|AB|^2 = |A|^2|B|^2$.

(These and some other details concerning S will be left for the exercises at the end of this section.)

Then $|A|^2$ is a nonnegative integer for A in S, and if $C \neq 0, \pm 1$ is not prime in S, then $C = AB$, where

$$1 < |A|^2 < |C|^2 \qquad \text{and} \qquad 1 < |B|^2 < |C|^2.$$

From this one can show using induction II on $|C|^2$ that any element $|C|$ of S other than 0, 1, and -1 has a prime factorization.

We will show that 2 is prime in S. Suppose $2 = AB$, where neither A nor B is 1 or -1. Then

$$|A|^2|B|^2 = |AB|^2 = |2|^2 = 4,$$

by property (3). Since $|A|^2$ and $|B|^2$ are both ordinary integers > 1, we must have $|A|^2 = 2$. Let $A = a + b\sqrt{-6}$. Then $a^2 + 6b^2 = 2$. This is impossible whether $b = 0$ or b is not 0.

A similar proof (depending on the fact that $a^2+6b^2 = 5$ is impossible) shows that 5 is prime in S. Another prime in S is $2+\sqrt{-6}$. For suppose $2+\sqrt{-6} = AB$, with $1 < |A|^2$ and $1 < |B|^2$. Then

$$|A|^2|B|^2 = |AB|^2 = |2 + \sqrt{-6}|^2 = 2^2 + 6 \cdot 1^2 = 10.$$

By the assumptions on $|A|^2$ and $|B|^2$ we must have $|A|^2 = 2$ or $|A|^2 = 5$. But both of these are impossible by our previous calculations. A similar proof shows that $2 - \sqrt{-6}$ is prime in S.

Now observe that $(2 + \sqrt{-6})(2 - \sqrt{-6}) = 2^2 - (\sqrt{-6})^2 = 4 - (-6) = 10$. Thus we have the two distinct factorizations

$$2 \cdot 5 = (2 + \sqrt{-6})(2 - \sqrt{-6})$$

of 10 into primes in S.

Dividing Exactly

Now we return to the ordinary integers. Since the factorization of an integer into primes is unique except for order, the number of times a particular prime appears is determined, justifying the following definition.

DEFINITION. divides exactly

Suppose p is a prime and $k > 0$. We say p^k **divides** a **exactly**, and write $p^k \parallel a$, in case $p^k \mid a$ but $p^{k+1} \nmid a$.

Examples. $3 \parallel 6$, $2^5 \parallel 96$, $9 \parallel 27$ is false, $4 \parallel 24$ is false, $4 \parallel 62$ is false. ◇

Problems for Section 2.3.

Problems 1 through 20 refer to the set S defined in this section.

1. Find $|7 + 3\sqrt{-6}|^2$.
2. Find $|3 - 7\sqrt{-6}|^2$.

In the next four problems, find B and write it in the form $a + b\sqrt{-6}$.

3. $(3 - 7\sqrt{-6})B = 147 - 40\sqrt{-6}$
4. $-200 + 19\sqrt{-6} = (2 + 3\sqrt{-6})B$
5. $(-7 + 2\sqrt{-6})B = 73$
6. $(3 + \sqrt{-6})B = -15$

In the next four problems, the number A given is not prime. List all divisors $B = a + b\sqrt{-6}$ of A with $1 < |B|^2 < |A|^2$ and a and b positive.

7. $A = 127$ 8. $A = 193$ 9. $A = 79$ 10. $A = 103$

B

11. Show that $|A|^2 = 0$ if and only if $A = 0$, and $|A|^2 = 1$ if and only if A is 1 or -1.
12. Suppose P is prime in S. How many divisors does P have?
13. Is 7 prime in S? List all its divisors.
14. Is 31 prime in S? List all its divisors.
15. Show that S is closed under addition.
16. Show that if A divides 1, then A is 1 or -1.
17. Show that 5 is prime in S.
18. Show that $2 - \sqrt{-6}$ is prime in S.
19. Show that 3 is prime in S.
20. Show that $3 + \sqrt{-6}$ is prime in S.

True-False. In the next 11 problems, tell which statements are true and give counterexamples for the others. Assume p is prime, a and b are nonzero integers, i, j, and k are positive integers, and $i < j$.

21. If $p^i \parallel a$ and $p^j \parallel b$, then $p^{i+j} \parallel ab$.
22. If $p^i \parallel (a, b)$, then $p^{2i} \parallel ab$.
23. If $p^i \parallel a$ and $(a, b) = 1$, then $p^i \parallel ab$.
24. If $p^i \parallel a$ and $(a, b) > 1$, then $p^i \parallel ab$ is false.
25. If $p^j \parallel a$ and $p^i \parallel (a, b)$, then $p^{j+i} \parallel ab$.
26. If $p^i \parallel a$ and $p^j \parallel (a, b)$, then $a + b$ is prime.
27. If $p^i \parallel a$, then $p^{ki} \parallel a^k$.
28. $(a^k, b^k) = (a, b)^k$.
29. If $p^i \parallel a$ and $p^i \parallel b$, then $p^i \parallel a + b$.
30. If $p^i \parallel a$ and $p^j \parallel b$, then $p^i \parallel a + b$.
31. If $p^i \parallel a$ and $p^i \parallel b$, then there exists $m \geq i$ such that $p^m \parallel a + b$.

32. Give a solution to $x^2 + y^2 = z^2$ in positive integers other than the two given in this section.
33. When 2^{33} is computed on a certain hand calculator the answer 8589934605 is given. What does this have to do with the fundamental theorem of arithmetic?
34. Write $81 + 175\sqrt{-6}$ as a product of primes in S.

C

The next three problems refer to the set S.

35. Show that $|AB|^2 = |A|^2|B|^2$ in S.
36. Prove that if A in S is not 0, 1, or -1, then A has a prime factorization. Use induction II on $n = |A|^2$.
37. Show that if $|A|^2$ is prime in the ordinary sense, then A is prime in S.
38. Why is the concept of dividing exactly only applied to powers of primes?
39. Show that if $x^2 + y^2 = z^2$ and if no prime divides all three of x, y, and z, then z is odd.

40. Show that if it were proved that $x^n + y^n = z^n$ was impossible in positive integers for $n = 4$ and for n any odd prime, then Fermat's last theorem would be proved.
41. Very little is known of the life of Diophantus. A collection of problems published in the fifth or sixth century contains the following problem about him:

    "This tomb holds Diophantus. Ah, what a marvel! And the tomb tells scientifically the measure of his life. God vouchsafed that he should be a boy for the sixth part of his life; when a twelfth was added, his cheeks acquired a beard; He kindled for him the light of marriage after a seventh, and in the fifth year after his marriage He granted him a son. Alas! late-begotten and miserable child, when he had reached the measure of half his father's life, the chill grave took him. After consoling his grief by this science of numbers for four years, he reached the end of his life."

    How many years did Diophantus live?
42. Show that if $a^2 + b^2 = c^2$, then $3 \mid ab$.
43. Prove that if $n = 2^k$, $k > 0$, then $2^{k+2} \parallel 3^n - 1$.
44. Prove that if $n = 2^k t$, $k > 0$, $t > 0$ odd, then $2^{k+2} \parallel 3^n - 1$.
45. Prove that if a and b are positive integers, p is prime, and $a + b = 2p - 1$, then $p \parallel a!b!$.

2.4 Prime Power Factorizations

An Application of the Fundamental Theorem

The existence and uniqueness of the prime decomposition of an integer are very important for understanding the integers and proving things about them. Thinking about an integer in terms of its prime factorization provides an entry to many number-theoretic problems. When a proof using prime factorizations works, it is usually straightforward, if not necessarily elegant. The next theorem shows how the concepts of divisibility and of the gcd and lcm depend on prime factorizations.

Theorem 2.6. *Suppose a and b are positive integers. Let the distinct primes dividing a or b (or both) be $p_1, p_2, \ldots, p_n$. Suppose*

$$a = p_1^{j_1} p_2^{j_2} \cdots p_n^{j_n} \text{ and } b = p_1^{k_1} p_2^{k_2} \cdots p_n^{k_n},$$

where some of the exponents may be 0. Let m_i be the smaller and M_i the larger of j_i and k_i for $i = 1, 2, \ldots, n$. Then

1. *$a \mid b$ if and only if $j_i \leq k_i$ for $i = 1, 2, \ldots, n$.*
2. *$(a, b) = p_1^{m_1} p_2^{m_2} \cdots p_n^{m_n}$.*
3. *$[a, b] = p_1^{M_1} p_2^{M_2} \cdots p_n^{M_n}$.*

Proof. (Partial)

1. Suppose $a \mid b$. Then $b = ac$ for some integer c. Consider the prime factorizations of a, b, and c. According to the equation $b = ac$ every prime appearing in the factorization of a must also appear in that of b, and at least as many times. (Here we have used the fact that the factorization is unique up to order.) Thus $j_i \leq k_i$ for all i.

   Since part (1) contains an "if and only if" statement, this is only half of its proof. We leave the other half for the exercises at the end of this section.

2. Since $m_i \leq$ both j_i and k_i for all i, part (1) shows the expression on the right side of (2) to be a common divisor of a and b.

   Now suppose d is any common divisor of a and b. Then by part (1) again $d = p_1^{r_1} p_2^{r_2} \cdots p_n^{r_n}$, where $r_i \leq j_i$ and $r_i \leq k_i$ for all i. Then $r_i \leq m_i$ for all i and so d is less than or equal to the right side of (2). Thus the right side of (2) is a common divisor of a and b exceeding any other common divisor. By definition it is (a, b).

3. This proof is left for the problems.

□

Examples. [of the use of Theorem 2.6] Suppose $a = 75 = 3^1 5^2$, $b = 900 = 2^2 3^2 5^2$, and $c = 440 = 2^3 5^1 11^1$. Then

1. $a \mid b$, since $a = 2^0 3^1 5^2$, and $0 \leq 2$, $1 \leq 2$, and $2 \leq 2$;
2. $(b, c) = 2^2 3^0 5^1 11^0 = 20$;
3. $[b, c] = 2^3 3^2 5^2 11^1 = 19{,}800$.

Factoring Large Numbers

Using part (2) of the last theorem appears to be easier than using the Euclidean algorithm. This method depends on knowing the prime factorizations of the numbers involved, however. If n has mostly small prime factors, the factorization of n may be found by dividing through by each factor as it is found. If n has only large prime divisors, however, finding them may be difficult. In such a case the Euclidean algorithm and the equation $[a, b] = |ab|/(a, b)$ will still provide the easiest way to find the gcd and lcm of two numbers.

Very often the digits of a number tell us something about its divisors. For example, we know the number 7,586,634 is even without bothering to divide it by 2—because its last digit is even. This idea is so familiar that we have probably never thought about why it is true. Writing out a formal proof may be instructive.

Proposition D. *A positive integer written to base* 10 *is even if and only if its last digit is even.*

Proof. Suppose n has the digits $a_k, a_{k-1}, \ldots, a_0$, starting from the left, so

$$n = a_k 10^k + \cdots + a_1 10^1 + a_0.$$

We want to show that n and a_0 are either both even or both odd. Another way to say this is that

$$n \equiv a_0 \pmod{2}.$$

But this is implied by the fact that $10 \equiv 0 \pmod{2}$ and what we know about manipulating congruences (Theorems 1.16 and 1.21). □

Properties of Digits

The key to the proof just given is the fact that $10 \equiv 0 \pmod{2}$, and we write numbers in terms of powers of 10 (no doubt because humans have 10 fingers). If a different base were used the theorem might not be true.

Similar theorems can be proved for other divisors d, provided that 10 is sufficiently simple modulo d. For example, the same proof shows that a number is divisible by 5 if and only if its last digit is. Actually, the proof says even more, namely, that a positive integer is congruent modulo 5 to its last digit. Thus we can say not only that 38,707 is not divisible by 5, but also that its least residue modulo 5 is 2, since this is true for 7.

A similar proof shows that a positive integer is congruent to the sum of its digits modulo both 3 and 9. In particular, n is divisible by 3 (or 9) if and only if the sum of its digits is. We will do the proof only for 3.

Proposition E. *Any positive integer is congruent to the sum of its digits modulo 3.*

Proof. Let $n = a_k 10^k + \cdots + a_0$. Note that $10 \equiv 1 \pmod 3$. Then

$$n \equiv a_k(1)^k + \cdots + a_1(1) + a_0 \pmod 3.$$

□

A similar theorem holds for the modulus 11. We omit the proof.

Proposition F. *If* $n = a_k 10^k + \cdots + a_0$, *then*

$$n \equiv a_0 - a_1 + a_2 - \cdots \pmod{11}.$$

The following summarizes our results on digits.

Theorem 2.7 (the digit theorem). *Let* $n > 0$ *have the decimal representation* $a_k a_{k-1} \cdots a_0$. *Then*

1. $n \equiv a_0 \pmod 2$
2. $n \equiv a_0 \pmod 5$
3. $n \equiv a_k + \cdots + a_0 \pmod 3$
4. $n \equiv a_k + \cdots + a_0 \pmod 9$
5. $n \equiv a_0 - a_1 + \cdots \pmod{11}$.

Example. We will factor $n = 37{,}719$. Since $3 + 7 + 7 + 1 + 9 = 27$, we see $9 \mid n$. Division produces $n = 9 \cdot 4191$. Now we concentrate on 4191. Since $4 + 1 + 9 + 1 = 15$, 4191 is divisible by 3 but not 9. We have $4191 = 3 \cdot 1397$. Ordinary division shows $7 \nmid 1397$. The next prime is 11. Since $7 - 9 + 3 - 1 = 0$, 1397 is divisible by 11. This gives $1397 = 11 \cdot 127$.

Now $13^2 = 169$, so $127/13 < 13$. Thus if 127 had a proper factor ≥ 13, then it would also have a factor < 13. Since the latter possibility has been eliminated, we conclude that 127 is a prime. Thus $37{,}719 = 3^3 \cdot 11 \cdot 127$. ◇

Theorem 2.8. *If the integer* $n > 1$ *has no prime divisor* $\leq \sqrt{n}$, *then* n *is prime.*

Proof. This proof is illustrated by the argument about 13 at the end of the last example. Suppose n is composite. Then $n = d_1 d_2$, where both d_1 and d_2 exceed 1. If both d_1 and d_2 exceed $\sqrt{n}$, then

$$n = d_1 d_2 > (\sqrt{n})^2 = n,$$

which is impossible. Suppose $d_1 \leq \sqrt{n}$. Then d_1 is either prime or else has a prime divisor $\leq \sqrt{n}$. □

Problems for Section 2.4.

A

In each of the first 10 problems, write each integer in the form $p_1^{k_1} p_2^{k_2} \cdots p_n^{k_n}$, *where the ps are distinct primes, the ks are positive integers, and* $p_1 < p_2 < \cdots < p_n$. *This is called the* **prime power factorization** *of the integer.*

1. 293
2. 1001
3. 1763
4. 2310
5. 6561
6. 1,000,000
7. 1111
8. 16,281
9. 14,641
10. 44,032

Find the largest power of 3 *dividing* n *in each of the next four problems.*

11. $n = 2139$
12. $n = 142{,}155$
13. $n = 57{,}035$
14. $n = 11{,}799$

In each of the next six problems, use the prime factorization of a *and* b *(1) to decide whether* $a \mid b$, *(2) to compute* (a, b), *and (3) to compute* $[a, b]$.

15. $a = 495$, $b = 3861$
16. $a = 4600$, $b = 2116$
17. $a = 266$, $b = 748$
18. $a = 1287$, $b = 9009$
19. $a = 2090$, $b = 1911$
20. $a = 2592$, $b = -3888$

B

In the next four problems, suppose that (to base 10) n *consists of the digit* d *repeated* k *times. If* $d = 3$ *and* $k = 5$, *for example, then* $n = 33{,}333$.

21. When does 11 divide n?
22. When does 3 divide n?
23. When does 9 divide n?
24. When does 2 divide n?
25. Prove that any positive integer is congruent to its last digit modulo 5.
26. Prove that any positive integer is congruent to the sum of its digits modulo 9.
27. What is the least residue of 94,850,372 modulo 9? Modulo 11?
28. What is the least residue of 1,280,054 modulo 9? Modulo 11?
29. What is the least residue of $-85{,}763$ modulo 3? Modulo 11?
30. What is the least residue of $-158{,}473$ modulo 5? Modulo 11?
31. What is the least residue of 10^{99} modulo 9? Modulo 11?
32. What is the least residue of $17{,}615^3$ modulo 11?
33. What is the least residue of $65{,}432^3$ modulo 9?
34. What is the least positive integer k that makes the following statement true: "If $1 < n < 1000$ and if n has no positive divisor d with $1 < d < k$, then n is prime."?
35. What is the least positive integer k that makes the following statement true: "If $1 < n < 2000$ and if n has no positive divisor d with $1 < d < k$, then n is prime."?
36. Prove part (5) of the digit theorem.
37. Show that in the notation of the digit theorem $n \equiv 10a_1 + a_0 \pmod 4$.
38. A calculator says $2^{31} = 2147483654$. Show this is wrong.
39. How many ways can the digits 2, 3, 4, 5, 6 be arranged to get a number (such as 56,243) divisible by 11?

40. How many ways can the digits 1, 2, 3, 4, 5 be arranged to get a number divisible by 11?

C

41. Give a test for the divisibility of n by 3 depending on the digits of n when written to base 9.
42. Give a test for the divisibility of n by 2 depending on the digits of n when written to base 9.
43. Show that if n is a positive integer, then n is divisible by 2^k if and only if the number formed by dropping all but the last k digits of n is also.
44. Give the rest of the proof for part (1) of Theorem 2.6.
45. Prove part (3) of Theorem 2.6.
46. Give a proof of Theorem 1.13 using the fundamental theorem of arithmetic.
47. Show that a and b are relatively prime if and only if whenever p is prime and $p \mid a$, then $p \nmid b$.
48. Give a proof of Theorem 1.5 using Theorem 2.6.
49. Prove that if n is a positive integer and n^2 and $2n^2$ are written in base 3, then their last nonzero digits are 1 and 2, respectively.
50. Use the last problem to prove that $\sqrt{2}$ is irrational.
51. Show that $111 = k^2$ cannot be correct in any base.
52. Suppose $n > 0$ has the property that whenever $p \mid n$, p prime, then $p^2 \mid n$. Show that there exist integers a and b such that $n = a^3b^2$.

2.5 The Set of Primes is Infinite

The Sieve of Eratosthenes

Suppose we want to determine all the primes less than 100. We might proceed as follows. First we write out the integers from 2 to 100. We know 2 is prime; let us circle it and cross out the remaining even numbers on our list. Now the lowest number that hasn't been crossed out is 3; we circle it and cross out every third number thereafter, since 3 is a proper divisor of each of these. Our list now starts as follows:

② ③ ~~4~~ 5 ~~6~~ 7 ~~8~~ ~~9~~ ~~10~~ 11 ~~12~~ 13 ~~14~~ ~~15~~ ~~16~~ 17...

At each stage in this procedure the smallest number that has not been circled or crossed out must be prime, since otherwise it would have a smaller prime divisor, and so have been eliminated already.

This process is called the **sieve of Eratosthenes,** after the Greek scientist who invented it. Note that according to Theorem 2.8, in order to find all the primes up to n we need only sieve out multiples of primes $\leq \sqrt{n}$. To find the primes up to 100, for example, we need only cross out multiples of 2, 3, 5, and 7.

The operation of the sieve of Eratosthenes suggests that primes should be rarer among the larger integers. For example, its application to the numbers between 100 and 150 consists in crossing out the multiples of the 5 primes 2, 3, 5, 7, and 11 not exceeding $\sqrt{150} \approx 12.2$. Applying the sieve between 1000 and 1050, however, we eliminate not only the multiples of all these primes, but also the multiples of the additional 6 primes 13, 17, 19, 23, 29, and 31 between 12.2 and $\sqrt{1050} \approx 32.4$. It turns out that there are 10 primes between 100 and 150, but only 8 between 1000 and 1050. Between 10,000 and 10,050 there are just 4 primes.

These considerations suggest the possibility that at some stage in the application of the sieve all larger numbers will have been crossed out, so there would be no more primes. This would mean that there would exist only finitely many primes, say $p_1, p_2, \ldots, p_k$, so that each integer greater than 1 could be written as a product of powers of these primes. The integer

$$P = p_1 p_2 \cdots p_k$$

would then have the interesting property that $(P, n) > 1$ whenever $n > 1$, since all the distinct prime factors of n would appear in P. Let us make a table of (P, n) for small values of n.

n	1	2	3	4	5	6	7	8	9	10
(P,n)	1	2	3	2	5	6	7	2	3	10

Back in Section 1.5 we made a similar table for $(6, n)$. It turned out to be periodic, with period 6. In fact we proved in Theorem 1.19 that if $b \equiv b'$

(mod a), then $(a, b) = (a, b')$. But this doesn't square with what we know about (P, n). Theorem 1.19 says that if $n \equiv 1 \pmod{P}$, then $(P, n) = (P, 1)$; for example $(P, P+1) = 1$. Our definition of P, however, led us to the conclusion that $(P, n) > 1$ for $n > 1$. The trouble must be in our assumption that the number of primes is finite, which has produced a contradiction. Thus we have proved the following theorem:

Theorem 2.9. *The number of primes is infinite.*

Euclid's Proof That the Number of Primes is Infinite

Theorem 2.9 first appears in the works of Euclid, so we will give his proof, which has the advantage of depending on little beyond the definition of the primes.

As before, we assume there are only the primes $p_1, p_2, \ldots, p_k$ and no more. Consider the number

$$Q = p_1 p_2 \cdots p_k + 1.$$

Now Q is either prime or else has a prime factor. If Q is prime, we have a contradiction, since $p_1, p_2, \ldots, p_k$ are supposed to be all the primes. But any prime factor of Q must be different from all of p_1, p_2, $\ldots$, p_k, since it is easy to see that none of the ps can divide Q. This again contradicts the assumption that p_1, p_2, $\ldots$, p_k comprise all the primes.

Euler's Proof That the Number of Primes is Infinite

Now we give a more sophisticated proof of Theorem 2.9 by the Swiss mathematician Leonhard Euler. Some exposure to infinite series is desirable (but not absolutely necessary) for its understanding. Again we assume p_1, p_2, $\ldots$, p_k are all the primes. Then if n is any positive integer we may choose r big enough so that all the terms 1, 1/2, 1/3, $\ldots$, $1/n$ appear when we multiply out the product

$$\left(1+\frac{1}{p_1}+\frac{1}{p_1^2}+\cdots+\frac{1}{p_1^r}\right)\left(1+\frac{1}{p_2}+\frac{1}{p_2^2}+\cdots+\frac{1}{p_2^r}\right)\cdots\left(1+\frac{1}{p_k}+\frac{1}{p_k^2}+\cdots+\frac{1}{p_k^r}\right).$$

In fact increasing r merely adds in more terms. For example we get 1/12 by choosing $1/2^2$ from the first factor, 1/3 from the second, and 1 from all the others (assuming that $p_1 = 2$ and $p_2 = 3$).

Now by the formula for the sum of a geometric progression (Theorem 1.23)

$$1+\frac{1}{p}+\frac{1}{p^2}+\cdots+\frac{1}{p^r} = \frac{1-(p^{-1})^{r+1}}{1-p^{-1}} < \frac{1}{1-p^{-1}} = \frac{p}{p-1}.$$

We see that for any n the sum

$$1+\frac{1}{2}+\frac{1}{3}+\cdots+\frac{1}{n}$$

cannot exceed

$$\frac{p_1}{p_1 - 1} \cdot \frac{p_2}{p_2 - 1} \cdot \cdots \cdot \frac{p_k}{p_k - 1}.$$

But this contradicts the divergence of the harmonic series $1+1/2+1/3+\cdots$. (To get the contradiction without knowing about the harmonic series, see problem 20 at the end of this section.)

Other Theorems Concerning Primes

The distribution of primes has been studied extensively and is a central topic in what is known as **analytic number theory**. Unfortunately what can be proved mostly involves techniques too advanced to be presented here. The **prime number theorem**, which was proved independently in 1896 by Hadamard and de la Vallée-Poussin, states that

$$\frac{\pi(x)}{x/\log_e x} \to 1 \quad \text{as} \quad x \to \infty,$$

where $\pi(x)$ is the number of primes $\leq x$.

There is a scarcity of results implying the existence of infinitely many primes of special forms. For example primes p and q and called **twin primes** if they differ by 2. Examples are 3 and 5, 5 and 7, 11 and 13, 41 and 43, and 1,000,000,009,649 and 1,000,000,009,651. When a list of large primes is compiled, twin primes appear to continue to pop up no matter how far out one goes, but the **twin prime conjecture**, which states that there are infinitely many of them, has never been proved.

The one positive result of this type is **Dirichlet's theorem**, which says that if a and b are relatively prime, then the infinite arithmetic progression a, $a+b$, $a+2b$, ... contains infinitely many primes.

Factorization by Computer

Before the development of electronic computers mechanical devices were constructed to perform tedious number-theoretic computations. Figure 2.2 shows a photograph of a machine made by D. H. Lehmer of the University of California. Such devices played an important part in primality testing and factorization before the birth of modern computers.

Problems for Section 2.5.

A

Use the sieve of Eratosthenes to do each of the first four problems.

1. Find all primes between 50 and 100.
2. Find all primes between 100 and 150.
3. Find all primes between 1000 and 1025.

Figure 2.2 D. H. Lehmer's factoring machine

4. Find all primes between 1025 and 1050.

In the next two problems, pretend the number P used in the first proof of Theorem 2.9 really exists.

5. Find (P, n) for $n = 11, 12, \ldots, 20$.
6. Find (P, n) for $n = 21, 22, \ldots, 30$.

B

7. Define a legitimate function $P(n)$ to take the place of the fictional (P, n).
8. Give a detailed proof of why the number $p_1 p_2 \cdots p_k + 1$ of Euclid's proof of Theorem 2.9 is not divisible by any of the ps.
9. How many primes are congruent to 0 (mod 4)? To 2 (mod 4)?
10. Show that if $(a, b) > 1$, then there exists at most one prime congruent to $a \pmod b$.
11. Show that if p and q are twin primes and $p + q > 8$, then 12 divides $p + q$.
12. Show that if p, $p + 2$, and $p + 4$ are all prime, then $p = 3$.
13. Show that if $ab = c^2$, with a and b relatively prime positive integers, then $a = r^2$ and $b = s^2$ for some integers r and s.

C

14. Let S be a finite set of positive integers. Show, without using Theorem 2.9, that there exists $n > 1$ such that $(n, x) = 1$ for all x in S.
15. Disprove the statement: If p_i is the ith prime, then $p_1 p_2 \cdots p_k + 1$ is prime for all positive integers k.
16. Prove that there are infinitely many primes (such as 3, 7, 11, and 19) that are congruent to 3 (mod 4). (**Hint:** Suppose there are only $p_1, p_2, \ldots, p_k$. Show that $4p_1 p_2 \cdots p_k - 1$ must have a prime divisor $\equiv 3 \pmod 4$. See problem 9.)
17. Prove that there are infinitely many primes congruent to 5 (mod 6).
18. Prove, without using problem 16, that there exist infinitely many pairs of distinct primes p and q such that $p + q \equiv 2 \pmod 4$.
19. Show that if k is any positive integer, then the numbers $(k + 1)! + j$, $j = 2, 3, \ldots, k + 1$, are all composite. (Thus there exist arbitrarily long sequences of consecutive composite numbers.)
20. Show by induction on n that for all positive integers n

$$1 + \frac{1}{2} + \frac{1}{3} + \cdots + \frac{1}{2^n} > \frac{n+1}{2}.$$

Conclude that $1 + 1/2 + \cdots + 1/n$ can be made arbitrarily large.

The remaining problems are for those with some knowledge of infinite series.

21. Show that for k any positive integer

$$\left(1 + \frac{1}{p}\right)\left(1 + \frac{1}{p^2} + \frac{1}{p^4} + \cdots + \frac{1}{p^{2k}}\right) = 1 + \frac{1}{p} + \frac{1}{p^2} + \cdots + \frac{1}{p^{2k+1}}.$$

22. Let the ith prime be p_i, and let N be the set of all positive integers all of whose prime factors are $\leq p_k$. Use the last problem to show

$$\left(1+\frac{1}{p_1}\right)\left(1+\frac{1}{p_2}\right)\ldots\left(1+\frac{1}{p_k}\right)\sum_{n\in N}\frac{1}{n^2}=\sum_{n\in N}\frac{1}{n}.$$

23. Conclude from the last problem that

$$\left(1+\frac{1}{p_1}\right)\left(1+\frac{1}{p_2}\right)\ldots\left(1+\frac{1}{p_n}\right)\to\infty \quad \text{as} \quad n\to\infty.$$

24. Show that if $C>0$, then $e^C>1+C$.
25. Use the last two problems to show that $1/p_1+1/p_2+\ldots$ diverges. (**Hint:** Let $C=1/p_i$.)
26. Let S be a finite set of positive integers, and let T be the set of all finite products of elements of S. Show that $\sum 1/n$ converges, where the sum is taken over T.
27. Show that if $f(x)$ is a polynomial of degree at least 2, then the set of values of $f(n)$, $n=1,2,\ldots$, does not contain every prime.
28. Show that if p_i is the ith prime, then

$$\left(1-\frac{1}{p_1}\right)\left(1-\frac{1}{p_2}\right)\ldots\left(1-\frac{1}{p_n}\right)\to 0 \quad \text{as} \quad n\to\infty.$$

2.6 A Formula for $\tau(n)$

We now return to the evaluation of $\tau(n)$, which was sidetracked when we got into the much more interesting problem of prime factorization. What we know now makes the problem easy. Suppose

$$n = p_1^{k_1} p_2^{k_2} \dots p_t^{k_t},$$

where the ps are distinct primes. If d is a positive divisor of n, then by the first part of Theorem 2.6

$$d = p_1^{j_1} p_2^{j_2} \dots p_t^{j_t},$$

where $0 \le j_i \le k_i$ for $i = 1, 2, \dots, t$. (We allow the js to equal 0 to take care of primes not dividing d at all.)

For example, if $n = 63 = 3^2 7$, then the positive divisors of n are

$$\begin{array}{ccc} 3^0 7^0 & 3^1 7^0 & 3^2 7^0 \\ 3^0 7^1 & 3^1 7^1 & 3^2 7^1. \end{array}$$

Returning to the general case, we note that the unique factorization theorem also tells us that each different choice for j_1 through j_t gives us a different d. Since there are $k_1 + 1$ possibilities for j_1 (namely $0, 1, \dots, k_1$), $k_2 + 1$ possibilities for j_2, etc., we have the following theorem:

Theorem 2.10. *If $n = p_1^{k_1} p_2^{k_2} \dots p_t^{k_t}$, where $p_1, p_2, \dots, p_t$ are distinct primes, then*

$$\tau(n) = (k_1 + 1)(k_2 + 1) \dots (k_t + 1).$$

Example. Since $63 = 3^2 7$, we have $\tau(63) = (2 + 1)(1 + 1) = 6$. Likewise $\tau(120) = \tau(2^3 3^1 5^1) = (3 + 1)(1 + 1)(1 + 1) = 16$. ◇

Writing Divisors in a Multiplication Table

When, before the theorem, we wrote out the positive divisors of 63, we found it convenient to organize them into a rectangular array. This array may have reminded the reader of a multiplication table; indeed, that is exactly what it was. Let us look at it again.

	1	3	9
1	1	3	9
7	7	21	63

We have written the positive divisors of 9 along the top and those of 7 along the left side. Each entry in the table is the product of a divisor of 9 and one of 7. Thus in this case

$$\tau(63) = \tau(9)\tau(7) = 3 \cdot 2 = 6.$$

This suggests that we might prove in general that $\tau(ab) = \tau(a)\tau(b)$ by showing that the divisors of ab are just all products of a divisor of a with one of b.

Let us suppose a and b have the positive divisors $d_1, d_2, \ldots, d_s$ and $e_1, \ldots, e_t$, respectively. Thus $\tau(a) = s$ and $\tau(b) = t$. We consider the array

$$\begin{array}{cccc} d_1e_1 & d_2e_1 & \ldots & d_se_1 \\ d_1e_2 & d_2e_2 & \ldots & d_se_2 \\ & & \vdots & \\ d_1e_t & d_2e_t & \ldots & d_se_t. \end{array}$$

This array has st entries, so in order to show $\tau(ab) = st$ we must demonstrate three things:

1. Every entry is a divisor of ab.

2. Every positive divisor of ab appears in the array.

3. No two entries are equal.

Since (1) is easy to see we proceed to (2). Suppose c is a positive divisor of ab. Then there exists an integer k such that $ab = ck$. Consider the prime factorizations of both sides of this equation. By the fundamental theorem of arithmetic we can match up the primes on both sides in a one-to-one fashion. In particular, some of the primes in the factorization of c match up with primes in the factorization of a, and the rest with primes in b. We see that c is a product of a divisor of a with one of b.

We now turn to condition (3). Suppose that two entries of the array are equal, say $d_ie_j = d_ke_l$. If we know that the prime factors of d_i could only occur in d_k, and vice versa, we could conclude that $d_i = d_k$, and $e_j = e_l$ would follow. Unfortunately, this need not be true. For example, if p is a prime dividing both a and b, then $p \cdot 1$ and $1 \cdot p$ will appear at different places in the array.

Evidently we need an additional hypothesis in order to prove (3), namely that no prime divides both a and b. If this is the case, we easily see that $d_ie_j = d_ke_l$ implies that $d_i = d_k$ and $e_j = e_l$, since, for example, the primes dividing d_i cannot appear in e_l, a divisor of b, and so must all turn up in d_k. Since saying no prime divides both a and b is equivalent to saying $(a, b) = 1$, we have proved the following theorem:

Theorem 2.11. *If a and b are relatively prime positive integers, then $\tau(ab) = \tau(a)\tau(b)$.*

The analysis above involves a more specific discovery that will be used again.

Theorem 2.12. *If a and b are relatively prime positive integers, then the set of positive divisors of ab consists exactly of all products de, where d is a positive divisor of a and e is a positive divisor of b. Furthermore, these products are all distinct.*

DEFINITION. multiplicative

A function $f(n)$ is said to be **multiplicative** if $f(ab) = f(a)f(b)$ whenever a and b are relatively prime positive integers.

Problems for Section 2.6.

A

In each of the first 10 problems, find $\tau(n)$ by means of the formula of this section.

1. $n = 75$
2. $n = 45$
3. $n = 30$
4. $n = 84$
5. $n = 1{,}000{,}000$
6. $n = 640$
7. $n = 900$
8. $n = 2310$
9. $n = 4961$
10. $n = 17{,}640$

In each of the next four problems, make a multiplication table of products of positive divisors of a times positive divisors of b. Are all the entries divisors of ab? Does every positive divisor of ab appear? Are the entries distinct?

11. $a = 8$, $b = 15$
12. $a = 15$, $b = 66$
13. $a = 28$, $b = 21$
14. $a = 21$, $b = 26$

15. Show that if $d \mid a$ and $e \mid b$, then $de \mid ab$.

B

In each of the next seven problems, show that the function f is multiplicative, that is, if $(a, b) = 1$, then $f(ab) = f(a)f(b)$. Is the hypothesis that $(a, b) = 1$ needed?

16. $f(n) = n$
17. $f(n) = 1/n$
18. $f(n) = 1$ if n is odd and $f(n) = 0$ if n is even
19. $f(n) = 1$ if $n = k^2$ for some integer k and $f(n) = 0$ otherwise
20. $f(n)$ is 1 or -1 according as n is odd or even
21. $f(n)$ is 0, -1, or 1 according as $4 \mid n$, $2 \parallel n$, or n is odd
22. $f(n) = 1$ if n is 1 or 4, $f(2) = -2$, and $f(n) = 0$ otherwise

True-False. In the next eight problems, tell which statements are true and give counterexamples for the others. Assume f and g are multiplicative functions, that $g(n) = 0$ for no n, that C is a real number, and that $k > 0$. We define the functions $f + g$, fg, f^k, Cf, f/g, and $f \circ g$ at n to be $f(n) + g(n)$, $f(n)g(n)$, $(f(n))^k$, $Cf(n)$, $f(n)/g(n)$, and $f(g(n))$, respectively.

23. $f + g$ is multiplicative.
24. fg is multiplicative.
25. Cf is multiplicative.
26. f^k is multiplicative.
27. f/g is multiplicative.
28. $f \circ g$ is multiplicative.
29. $f(1) = 1$.
30. $g(1) = 1$.

31. Find the smallest n such that $\tau(n) = 8$.
32. Find the smallest n such that $\tau(n) = 35$.
33. Find the smallest n such that $\tau(n) = 105$.

34. Find the smallest n such that $\tau(\tau(n)) = 5$.
35. Find x such that $\tau(10x) = 10$.
36. Find x such that $\tau(10x) = x$.
37. Show that if p is prime, then there exists x such that $\tau(px) = x$.
38. Show that if p is prime, then there exists x such that $\tau(p^2x) = x$.
39. Show that if p is prime, then there exists x such that $\tau(2px) = x$.
40. Prove that if exactly t distinct primes divide n, then $\tau(n^2) \leq 2^t\tau(n)$.

C

41. Prove that $\tau(n)$ is multiplicative by means of the formula in this section.
42. Prove that if $(a, b) > 1$ then $\tau(ab) < \tau(a)\tau(b)$.
43. Prove that $\tau(n)$ is odd if and only if n is a perfect square.
44. Show that if $m > 1$, then there exist infinitely many integers n such that $\tau(n) = m$.
45. Show that if $n = k^m$, then $\tau(n) \equiv 1 \pmod{m}$.
46. Show that if $\tau(n)$ is prime, then n is a power of a prime.
47. Show that if $n > 0$, then as d runs through the positive divisors of n, so does n/d.
48. Show that if $d \mid a$, $e \mid b$, and $(a, b) = 1$, then $(d, e) = 1$.
49. Show that if $n > 0$, then there exists x such that $\tau(nx) = n$.
50. Show that if $n > 0$, the number of solutions to $\tau(nx) = n$ is 1 if and only if $n = 1$, 4, or a prime, is finite if n is the product of distinct primes, and is infinite otherwise.
51. Find the smallest $n > 0$ such that $\tau(nx) = x$ has no solution.
52. Show that $\tau(\tau(n)) = 2$ if and only if $n = p^{q-1}$, where p and q are primes.

Leonhard Euler

1703–1783

"Prolific" is the word most often applied to the Swiss mathematician Euler, from whom gushed a steady flow of work from the age of 19 on, even though he was blind for the last 17 years of his life. He produced more than 500 books and papers, and also 13 children.

Euler is responsible for much of the mathematical notation and conventions used today, including $f(x)$ for a function and a, b, c for the sides and A, B, C for the opposite angles of a triangle. Mainly known for his work in analysis, he wrote a calculus textbook and introduced the present-day symbols for e, π, and i, although he originally used the latter symbol for what we would now write as ∞. What we would express as $\lim_{n\to\infty}(1+x/n)^n$, Euler would write $(1+x/i)^i$. He discovered a wonderful equation linking the most important mathematical constants: $e^{\pi i}+1=0$.

Among Euler's discoveries in number theory is the law of quadratic reciprocity, which connects the solvability of the congruences

$$x^2 \equiv p \pmod q \quad \text{and} \quad y^2 \equiv q \pmod p,$$

where p and q are distinct primes, although it remained for Gauss to provide the first proof. He is also credited with the invention of graph theory, which has flourished in the twentieth century, with his solution of the Königsberg bridge problem.

Most of Euler's life was spent in the courts of Frederick the Great of Prussia and Catherine the Great of Russia, in the company of such luminaries as Voltaire. His phenomenal memory (he knew Virgil's *Aeneid* by heart) helped him continue his mathematical researches after he became blind, when he dictated his papers to his sons.

LEONHARD EULER

Chapter 3

Numerical Functions

Functions like $\tau(n)$, the number of positive divisors of n, defined on the positive integers, are variously called numerical, arithmetic, or number-theoretic functions. They are treated in this chapter.

3.1 The Sum of the Divisors

DEFINITION. $\sigma(n)$

Let n be a positive integer. We define $\sigma(n)$ to be the sum of the positive divisors of n.

Examples. If $n = 7$ we have $\sigma(n) = 1 + 7 = 8$; for $n = 9$ we have $\sigma(n) = 1 + 3 + 9 = 13$; and for $n = 63$ we have

$$\begin{aligned} \sigma(n) &= 1+3+9 \\ &+ 7+21+63 = 104. \end{aligned}$$

◇

The Multiplication Table Again

The astute reader may have noticed that $\sigma(63) = 104 = \sigma(7)\sigma(9)$, and that in fact in adding up the positive divisors of 63 we have again created a multiplication table of the divisors of 7 times those of 9. We are seeing a repetition of the ideas leading to our proof that $\tau(n)$ was multiplicative. Indeed, Theorem 2.12 of the last section may be used to prove $\sigma(n)$ multiplicative just as it was used to prove $\tau(n)$ multiplicative. The reader may want to try to provide his or her own proof for the next theorem before reading the one given.

Theorem 3.1. *The function $\sigma(n)$ is multiplicative.*

Proof. Suppose a and b are relatively prime positive integers. Let the positive divisors of a be $d_1, d_2, \ldots, d_s$, and let those of b be $e_1, e_2, \ldots, e_t$. Then

$$\begin{aligned}\sigma(a)\sigma(b) &= (d_1 + d_2 + \cdots + d_s)(e_1 + e_2 + \cdots + e_t) \\ &= \quad d_1e_1 + d_2e_1 + \cdots + d_se_1 \\ &\quad + d_1e_2 + d_2e_2 + \cdots + d_se_2 \\ &\qquad \vdots \\ &\quad + d_1e_t + d_2e_t + \cdots + d_se_t.\end{aligned}$$

But by Theorem 2.12 the numbers in this sum are exactly the positive divisors of ab. Thus $\sigma(a)\sigma(b) = \sigma(ab)$. □

The Summation Notation

There is a notation for sums that avoids the sprawl (and possible ambiguity) of the ... s used above. The Greek letter Σ (capital sigma) signals the sum of terms, each of which depends on an integer that varies between two limits. For example the sum

$$d_1 + d_2 + \cdots + d_s$$

of interest in the last theorem would be expressed as

$$\sum_{i=1}^{s} d_i.$$

In general, if f is some function of an integral variable and $m \leq n$, by

$$\sum_{i=m}^{n} f(i)$$

we mean

$$f(m) + f(m+1) + f(m+2) + \cdots + f(n).$$

The variable i is called the **index of summation** and is similar to the variable of integration in a definite integral in that it really does not matter what letter is used. For example

$$\sum_{i=1}^{5} i \quad \text{and} \quad \sum_{j=1}^{5} j$$

both mean the same thing (namely, 15), and any other letter could be used in place of i or j so long as it had no previous meaning.

In number theory, it is common to extend the summation notation to situations in which the index of summation does not run through a set of consecutive integers. A description of the values the index is allowed to assume is simply written under the Σ. For example,

$$\sum_{\substack{p \text{ prime} \\ p<7}} p^2 = 4 + 9 + 25 = 38, \quad \text{and} \quad \sigma(n) = \sum_{\substack{d|n \\ d>0}} d.$$

To interpret the last summation correctly the reader must realize that d and not n is the index of summation. One clue to this is that the n appears outside the summation (in $\sigma(n)$), something an index of summation never does.

When summing over the divisors of a number we generally wish to consider those that are positive, so in order to avoid always writing $d > 0$ under the sigma we establish the following convention:

CONVENTION When a summation is over the divisors of a number, these divisors are restricted to be positive.

There are various rules for manipulating summations that follow from the usual rules of algebra. For example,

$$\sum_{i=1}^{n} Kf(i) = K \sum_{i=1}^{n} f(i),$$

because

$$Kf(1) + Kf(2) + \cdots + Kf(n) = K(f(1) + \cdots + f(n))$$

according to the distributive law. Similar rules will be found in the exercises and can be proved by reverting to the $\cdots$ notation.

One way to express the number of elements in a set is to sum 1 over that set. For example,

$$\tau(n) = \sum_{d|n} 1.$$

Evaluating Multiplicative Functions

Knowing that a function is multiplicative is valuable because it reduces the problem of determining a formula to evaluation at a power of a prime. Suppose, for example, $n = p_1^{k_1} p_2^{k_2} \cdots p_m^{k_m}$, where the ps are distinct primes, and suppose we know f is a multiplicative function. Then

$$\begin{aligned} f(n) &= f(p_1^{k_1} p_2^{k_2} \cdots p_m^{k_m}) = f(p_1^{k_1}) f(p_2^{k_2} \cdots p_m^{k_m}) \\ &= f(p_1^{k_1}) f(p_2^{k_2}) f(p_3^{k_3} \cdots p_m^{k_m}) \\ &= \qquad \cdots \\ &= f(p_1^{k_1}) f(p_2^{k_2}) \cdots f(p_m^{k_m}). \end{aligned}$$

Evaluating f at a prime power is generally simpler than at an arbitrary integer. For example,

$$\sigma(p^k) = 1 + p + p^2 + \cdots + p^k.$$

This is a geometric progression, and by the formula we developed in Section 1.6 its sum is

$$\sigma(p^k) = \frac{p^{k+1} - 1}{p - 1}. \tag{3.1}$$

We are now in a position to give a formula for the function $\sigma(n)$, but the notation we have would again involve an excessive use of ellipsis (three dots). We need a compact symbolism for products analogous to the sigma notation for sums.

The Product Notation

The symbol Π (capital Greek letter pi) is used to denote a product. Thus

$$\prod_{i=1}^{n} f(i) = f(1)f(2)\cdots f(n),$$

and, more generally, if $S(i)$ is a statement about i, by

$$\prod_{S(i)} f(i)$$

we mean the product of the numbers $f(i)$ over all values of i for which $S(i)$ is true.

Examples.

$$\prod_{i=1}^{3} i^2 = 1 \cdot 4 \cdot 9 = 36,$$

$$\prod_{\substack{p \text{ prime} \\ p<5}} \frac{1}{p} = \frac{1}{2} \cdot \frac{1}{3} = \frac{1}{6},$$

and

$$\prod_{i=1}^{n} i = n!.$$

The following theorem follows from the fact that σ is multiplicative and equation (3.1) above.

Theorem 3.2. *If $n = p_1^{k_1} p_2^{k_2} \dots p_m^{k_m}$, where the ps are distinct primes, then*

$$\sigma(n) = \prod_{i=1}^{m} \frac{p_i^{k_i+1} - 1}{p_i - 1}.$$

Examples. If $n = 63 = 7 \cdot 3^2$, then

$$\sigma(n) = \frac{7^2 - 1}{7 - 1} \cdot \frac{3^3 - 1}{3 - 1} = \left(\frac{48}{6}\right)\left(\frac{26}{2}\right) = 8 \cdot 13 = 104.$$

Likewise

$$\sigma(1{,}000{,}000) = \sigma(2^6 5^6) = \frac{2^7 - 1}{2 - 1} \cdot \frac{5^7 - 1}{5 - 1} = \left(\frac{127}{1}\right)\left(\frac{78{,}124}{4}\right) = 2{,}480{,}437.$$

◊

Problems for Section 3.1.

A

In the first 10 problems, compute $\sigma(n)$.

1. $n = 20$
2. $n = 30$
3. $n = 100$
4. $n = 81$
5. $n = 101$
6. $n = 40$
7. $n = 667$
8. $n = 1000$
9. $n = 272$
10. $n = 1331$

Define $s_2(n)$ to be $\sum_{d|n} d^2$. Compute $s_2(n)$ in each of the next four problems.

11. $n = 7$
12. $n = 9$
13. $n = 63$
14. $n = 100$

Define $s_{-1}(n)$ to be $\sum_{d|n}(1/d)$. Compute $s_{-1}(n)$ in each of the next four problems.

15. $n = 7$
16. $n = 9$
17. $n = 63$
18. $n = 28$

In the next 12 problems, compute each sum or product.

19. $\displaystyle\sum_{i=1}^{5} i^2$
20. $\displaystyle\sum_{i=3}^{6} \frac{1}{i}$
21. $\displaystyle\sum_{d|12} (d + 0.5)$
22. $\displaystyle\sum_{d|18} \frac{6}{d}$
23. $\displaystyle\sum_{d|28} 1$
24. $\displaystyle\sum_{i=100}^{200} 1$
25. $\displaystyle\sum_{\substack{|p-10|<5 \\ p \text{ prime}}} p$
26. $\displaystyle\sum_{\substack{p \text{ prime} \\ p-1 \text{ square} \\ p<20}} p$
27. $\displaystyle\prod_{i=3}^{5} \frac{i}{i+1}$

28. $\prod_{i=-2}^{1} i$ 29. $\prod_{d|10} (d+1)$ 30. $\prod_{\substack{p \text{ prime} \\ p|24}} p^2$

B

Prove the formula in each of the next four problems. All sums and products run from $i = 1$ to n.

31. $\sum K = nK$
32. $\sum(f(i) + g(i)) = \sum f(i) + \sum g(i)$
33. $\prod Kf(i) = K^n \prod f(i)$
34. $\prod(f(i))^K = (\prod f(i))^K$
35. Show by induction on n that

$$\sum_{i=1}^{n} Kf(i) = K \sum_{i=1}^{n} f(i),$$

using only that $K(a+b) = Ka + Kb$.

36. Suppose $m \leq n$. What is

$$\sum_{i=m}^{n} 1?$$

37. Give a noninduction proof of the formula for the sum of a geometric progression of Section 1.6 by letting the sum be S and showing that $RS - S = A(R^n - 1)$.

C

38. Show that the function s_2 defined above is multiplicative.
39. Show that the function s_{-1} defined above is multiplicative.
40. Show that if p is prime, then

$$s_2(p^k) = \frac{p^{2(k+1)} - 1}{p^2 - 1}.$$

41. Show that if p is prime, then

$$s_{-1}(p^k) = \frac{p^{k+1} - 1}{p^k(p-1)}.$$

42. Use Problem 47 of Section 2.6 to show that $n \cdot s_{-1}(n) = \sigma(n)$ for all n.
43. Use Problem 41 of this section to show that $n \cdot s_{-1}(n) = \sigma(n)$ for all n.
44. Show that if $\sigma(n)$ is odd, then $n = j^2 2^k$, where j is odd.
45. Show that

$$\left(\sum_{i=1}^{m} f(i)\right)\left(\sum_{j=1}^{n} g(j)\right) = \sum_{i=1}^{m}\left(\sum_{j=1}^{n} f(i)g(j)\right) = \sum_{j=1}^{n}\left(\sum_{i=1}^{m} f(i)g(j)\right).$$

46. Prove by induction that if $p_1, p_2, \ldots, p_k$ are distinct primes and if f is multiplicative, then

$$f(p_1^{m_1} p_2^{m_2} \cdots p_k^{m_k}) = f(p_1^{m_1}) f(p_2^{m_2}) \cdots f(p_k^{m_k}).$$

3.2 Multiplicative Functions

Manufacturing Multiplicative Functions

In the last two sections, we proved first $\tau(n)$ and then $\sigma(n)$ to be multiplicative, using almost the same proof. In the last problem set, the functions $s_2(n)$ and $s_{-1}(n)$ were defined to be respectively the sum of the squares and reciprocals of the positive divisors of n, and these both turned out to be multiplicative also. Let us see how far this method of proof can be carried.

Let us suppose that f is some arbitrary numerical function, and define the function F by

$$F(n) = \sum_{d|n} f(d).$$

As before, let us assume that a and b are relatively prime, with positive divisors $d_1, d_2, \ldots, d_s$ and $e_1, e_2, \ldots, e_t$, respectively. Then

$$\begin{aligned} F(a)F(b) &= (f(d_1) + \cdots + f(d_s))(f(e_1) + \cdots + f(e_t)) \\ &= \quad f(d_1)f(e_1) + \cdots + f(d_s)f(e_1) \\ &\quad + f(d_1)f(e_2) + \cdots + f(d_s)f(e_2) \\ &\qquad \vdots \\ &\quad + f(d_1)f(e_t) + \cdots + f(d_s)f(e_t). \end{aligned}$$

On the other hand, by Theorem 2.12 $F(ab)$ is

$$\begin{array}{ccc} f(d_1e_1)+ & \cdots & +f(d_se_1) \\ +f(d_1e_2)+ & \cdots & +f(d_se_2) \\ & \vdots & \\ +f(d_1e_t)+ & \cdots & +f(d_se_t). \end{array}$$

We see that in order for these two expressions to be equal, it is enough that $f(d)f(e) = f(de)$ whenever $d \mid a$ and $e \mid b$. Since in such circumstances d and e will be relatively prime if a and b are, it suffices that the function f be multiplicative in order that F be.

Theorem 3.3. *Suppose f is a multiplicative function, and define F by*

$$F(n) = \sum_{d|n} f(d).$$

Then F is multiplicative also.

Examples. Since the function defined by $f(n) = n$ for all n is easily seen to be multiplicative, then so is

$$\sigma(n) = \sum_{d|n} f(d) = \sum_{d|n} d.$$

In fact, if we define $f(n)$ to be n^k, we easily see f is multiplicative. Thus so is

$$s_k(n) = \sum_{d|n} f(d) = \sum_{d|n} d^k.$$

Thus at one swoop we have shown $\sigma(n)$ (taking $k = 1$), $\tau(n)$ (taking $k = 0$), and the functions of the problems in the last section (with $k = 2$ and -1) to be multiplicative. ♢

We now have a general scheme for inventing and finding formulas for multiplicative functions of the form

$$F(n) = \sum_{d|n} f(d).$$

We start with some simple function f that we prove to be multiplicative. Then by the last theorem F is also multiplicative. To find a formula for F (or for any multiplicative function) it suffices to evaluate F at prime powers, just as in our derivation of the formula for $\sigma(n)$ in Section 3.1.

As an example, we consider the function $g(n)$ defined to be 0 if n is even and 1 if n is odd. This is easily seen to be multiplicative since $g(ab)$ and $g(a)g(b)$ are both 1 if and only if a and b are both odd. Now let

$$G(n) = \sum_{d|n} g(d).$$

Then G is multiplicative by the last theorem. We evaluate G at prime powers. The prime 2 is a special case. We have

$$G(2^k) = g(1) + g(2) + \cdots + g(2^k) = 1 + 0 + 0 + \cdots + 0 = 1.$$

On the other hand, if p is an odd prime, then

$$G(p^k) = g(1) + g(p) + \cdots + g(p^k) = 1 + 1 + \cdots + 1 = k + 1.$$

We see that if $n = 2^k p_1^{k_1} p_2^{k_2} \dots p_s^{k_s}$, where the ps are distinct odd primes, then $G(n) = (k_1 + 1)(k_2 + 1)\dots(k_s + 1)$. This is similar to the formula for $\tau(n)$. In fact $G(n)$ is the number of *odd* divisors of n.

The Uniqueness of f

Since Theorem 3.3 is so convenient for proving a function to be multiplicative, we might wonder, given a function F, if we could find a function f such that for all positive integers n,

$$F(n) = \sum_{d|n} f(d).$$

This question will be settled in Section 3.6. For now we will content ourselves with showing that if such a function f exists, then it is unique, and in the process give another example of an induction II proof.

Theorem 3.4. *Suppose F is a numerical function. Then there is at most one function f such that*

$$F(n) = \sum_{d|n} f(d) \tag{3.2}$$

for all positive integers n.

Proof. Suppose that g is a function such that also

$$F(n) = \sum_{d|n} g(d) \tag{3.3}$$

for all positive integers n. We will show by induction II on n that $f(n) = g(n)$, $n = 1, 2, \ldots$.

1. Taking $n = 1$ in (3.2) and (3.3) we see that $F(1) = f(1)$ and $F(1) = g(1)$, so $f(n) = g(n)$ for $n = 1$.

2. Suppose for some integer k we have $f(n) = g(n)$ for $n = 1, 2, \ldots, k$. Then

$$F(k+1) = \sum_{d|k+1} f(d) = \sum_{d|k+1} g(d).$$

By the induction hypothesis the two sums on the right have all their terms equal except possibly $f(k+1)$ and $g(k+1)$, which must therefore also be equal.

Thus by the induction II principle we see that $f(n) = g(n)$ for all positive integers n. □

Why Study Number Theory?

Perhaps the reader is wondering what good a function like $\sigma(n)$ is. Who cares that we have found a formula for the sum of the positive divisors of n? This is a question that each person must answer for himself or herself, since number theory, like mountain climbing, is an endeavor practiced mainly just for its own sake. Among the reasons some people might care about it are:

Accomplishment. Solving a mathematical problem may give the same pleasant feeling of success as finishing a difficult crossword puzzle or winning a set of tennis.

Understanding. Knowing why a number is divisible by 9 if and only if the sum of its digits is (and thus the justification of the old check of "casting out 9s") may bring satisfaction.

Skill. A person might take pride in being able to play the guitar or water ski, or, just as well, count the divisors of a million in a few seconds or compute the greatest common divisor of two large numbers.

Beauty. Many find beauty in number theory, either in the structure of the integers themselves (as an astronomer might find beauty in the laws of the universe), or in an ingenious or elegant proof.

We do not want to leave the impression that there are no applications at all for number theory. The RSA method of public key cryptography, which is treated in Section 4.5, involves many number-theoretic concepts, and has gained a great deal of attention in recent years. Its security depends on the difficulty of factoring large numbers, a problem that has been studied for centuries, long before its connection to anything useful. This is similar to many episodes in science in which phenomena studied purely for their theoretical interest later turned out to have immense practical consequences (such as atomic energy).

The study of numbers for their own sake is my no means new. The Greeks were interested in numbers like 6, which is the sum of its positive divisors other than itself; $6 = 1 + 2 + 3$. If n is any number with this property, then $n = \sigma(n) - n$, since by definition $\sigma(n)$ adds in the divisor n itelf.

DEFINITION. perfect number

We say n is a **perfect number** in case $\sigma(n) = 2n$.

A Table for $\sigma(n)$

To look for perfect numbers other than 6 we will make a table of the first 30 values of $\sigma(n)$ as follows:

n	$\sigma(n)$	n	$\sigma(n)$	n	$\sigma(n)$
1	1	11	12	21	32
2	3	12	28	22	36
3	4	13	14	23	24
4	7	14	24	24	60
5	6	15	24	25	31
6	12	16	31	26	42
7	8	17	18	27	40
8	15	18	39	28	56
9	13	19	20	29	30
10	18	20	42	30	72

Construction of the table was simplified by using a few properties of $\sigma(n)$. The primes were easily filled in, since $\sigma(p) = p + 1$. Prime powers follow the rule that $\sigma(p^k) = \sigma(p^{k-1}) + p^k$. Thus $\sigma(4) = \sigma(2) + 4$, $\sigma(8) = \sigma(4) + 8$, etc. The remaining values were then determined by the fact that $\sigma(n)$ is multiplicative. For example, $\sigma(18) = \sigma(2)\sigma(9) = 39$. (But $\sigma(3)\sigma(6) = 48$. What's wrong?)

Interest in perfect numbers precedes Euclid, and many early writers made note of them. Saint Augustine said that God created the world in six days rather than all at once because "The perfection of the work is signified by the perfect number 6." Others asserted that there are infinitely many perfect numbers; that, indeed, there is exactly one between 1 and 10, another between 10 and

100, another between 100 and 1000, etc.; that all perfect numbers are even; and that the perfect numbers alternately end in the digits 6 and 8. None of these assertions has ever been proved, and some are known to be false. We will continue the study of perfect numbers in the next section.

DEFINITION. abundant and deficient numbers

A number is said to be **abundant** if the sum of the positive divisors of the number other than itself exceeds the number (so $\sigma(n) > 2n$), and **deficient** if this sum is less than the number (so $\sigma(n) < 2n$).

Examples. Since $\sigma(12) = 28 > 2 \cdot 12$, the number 12 is abundant. Since $\sigma(11) = 12 < 2 \cdot 11$, the number 11 is deficient. Clearly each positive integer is either perfect, abundant, or deficient. ♢

DEFINITION. amicable numbers

We say a pair of numbers a and b is **amicable** if the sum of the positive divisors of a less than a equals b, and the sum of the positive divisors of b less than b equals a. Another way to say this is that

$$\sigma(a) - a = b \qquad \text{and} \qquad \sigma(b) - b = a.$$

Example. The smallest pair of amicable numbers is 220 and 284. Here $220 = 2^2 5 \cdot 11$, and

$$\sigma(220) - 220 = \sigma(2^2)\sigma(5)\sigma(11) - 220 = 7 \cdot 6 \cdot 12 - 220 = 284.$$

Likewise $284 = 2^2 71$, and

$$\sigma(284) - 284 = \sigma(2^2)\sigma(71) - 284 = 7 \cdot 72 - 284 = 220.$$

♢

Amicable pairs have long been considered important in numerology (which has the same relation to number theory as astrology to astronomy). The following is from the writings of the Arab scholar Ibn Khaldun (1332–1406):

> Let us mention that the practice of the art of talismans has also made us recognize the marvelous virtues of amicable (or sympathetic) numbers. These numbers are 220 and 284. One calls them amicable because the aliquot parts of one when added give a sum equal to the other. Persons who occupy themselves with talismans assure that these numbers have a particular influence in establishing union and friendship between two individuals. One prepares a horoscope theme for each individual, the first under the sign of Venus while this planet is in its house or in its exaltation and while it presents in regard to the moon an aspect of love and benevolence. In the second theme the ascendant should be in the seventh sign. On each one of these themes one inscribes one of the numbers just indicated, but giving the strongest number to the person whose friendship one wishes to gain, the beloved person. I don't know if by the strongest number one wishes to designate the greatest one or the one which has

> the greatest number of aliquot parts. There results a bond so close between the two persons that they cannot be separated. The author of the Ghaia and other great masters in this art declare that they have seen this confirmed by experience.

The pair 220, 284 was the only one known to the ancients, and it was not until 1636 that another pair was found by Fermat. Now about 400 pairs of amicable numbers are known. In 1866 a sixteen-year-old Italian boy, Nicolo Paginini, found the pair 1184, 1210, which had been overlooked by mathematicians up to that time. Incredibly, only 220, 284 is smaller.

At a meeting of the American Mathematical Society in San Francisco in 1981 Hilton Chen and Dale Woods of Northeast Missouri State University announced two previously unpublished pairs of amicable numbers, the larger of which was

$$a = 21{,}741{,}269{,}040{,}875{,}890{,}083{,}566{,}772{,}572{,}567{,}935{,}979{,}368{,}836{,}363{,}843$$
$$b = 22{,}261{,}723{,}990{,}815{,}556{,}829{,}012{,}769{,}686{,}652{,}975{,}057{,}619{,}942{,}956{,}477.$$

Problems for Section 3.2.

1. Extend the table of $\sigma(n)$ to run from 31 to 50.
2. Extend the table of $\sigma(n)$ to run from 51 to 70.
3. Which of the numbers from 1 to 15 are abundant?
4. Which of the numbers from 16 to 30 are deficient?
5. Let $G(n)$ be the function defined in this section. Compute $G(n)$ for n from 1 to 10.
6. Compute $G(n)$ for n from 11 to 20.

We define the function $M(n)$ to be 1 if n is 1, 0 if $p^2 \mid n$ for any prime p, and $(-1)^k$ if $n = p_1p_2 \ldots p_k$, where the ps are distinct primes.

7. Compute $M(n)$ for $n = 1$ to 10.
8. Compute $M(n)$ for $n = 11$ to 20.
9. Check that 1184 and 1210 really are amicable.

B

In problems 10 and 11, assume f is a multiplicative function with the values given. Complete the table so far as you can be sure of the values.

10.

n	2	3	4	5	6	10
$f(n)$	3	4				8

11.

n	1	2	3	4	6	12
$f(n)$		4			5	

12. Prove that M (defined above) is multiplicative.
13. Show that if f and g are multiplicative functions, then so is fg, where $(fg)(n) = f(n)g(n)$.
14. Let $F(n) = \sum_{d \mid n} M(d)$. Compute $F(n)$ for $n = 1$, 3, 4, 6.
15. Let F be as in the last problem. Suppose p is prime and $k > 0$. Show $F(p^k) = 0$.

16. Let F be as in the last problem. Show $F(n)$ is 1 if $n = 1$ and 0 otherwise.
17. Let $F(n) = \sum_{e|n} \tau(e)$. Compute $F(n)$ for $n = 1$, 3, 4, and 6.
18. Let F be as in the last problem. Suppose p is prime. Show $F(p^k) = (k+1)(k+2)/2$.
19. Let F be as in the last problem and let $n = p_1^{m_1} p_2^{m_2} \cdots p_k^{m_k}$, the ps distinct primes. Give a formula for F.
20. Let $F(n) = \sum_{e|n} M(e)\tau(e)$. Give a formula for F.
21. Let $F(n) = \sum_{d|n} M(d)\sigma(d)$. Give a formula for F.
22. Show that all prime powers are deficient.
23. Show that a and b form an amicable pair if and only if $\sigma(a) = \sigma(b) = a+b$.
24. One of an amicable pair is 17,296. What is the other?

C

25. If $F(n) = \sum_{d|n} f(d)$ write $F = f'$. Show that $M'''(n) = \tau(n)$ for all n.
26. Let $m(n) = 0$ if $p^3 \mid n$, p prime. Otherwise let $m(n) = (-2)^k$, where k is the number of distinct primes p such that $p \parallel n$. Show that $m' = M$ in the notation of the last problem.
27. Prove that if f and g are multiplicative, then so is

$$F(n) = \sum_{d|n} f(d)g\left(\frac{n}{d}\right).$$

28. Show that if f is multiplicative and if $F(n) = \sum_{d|n} f(d)$ for all n, then

$$f(n) = \sum_{d|n} M(d)F\left(\frac{n}{d}\right)$$

for all n.

29. Let

$$F(n) = \frac{\tau(n)}{\sum_{e|n} 1/e}.$$

Show that $F(n) = n\tau(n)/\sigma(n)$ for all positive integers n.

3.3 Perfect Numbers

Looking for a Pattern

So far we have found only two perfect numbers: 6 and 28. Let us compute $\sigma(n)$ for each of these to try to see why it equals $2n$.

$$\sigma(6) = \sigma(2 \cdot 3) = \sigma(2)\sigma(3) = 3 \cdot 4$$

$$\sigma(28) = \sigma(4 \cdot 7) = \sigma(4)\sigma(7) = 7 \cdot 8$$

Each of the perfect numbers we know consists of a power of 2 times a prime. As the arrows indicate, when we apply σ to each, the power of 2 turns into the prime and the prime turns into twice the power of 2.

What must the relation between the prime and the power of two be for this to work? Suppose $n = 2^r p$, where p is prime. To get the same pattern we need

$$\sigma(2^r) = \frac{2^{r+1} - 1}{2 - 1} = p$$

and

$$\sigma(p) = p + 1 = 2 \cdot 2^r = 2^{r+1}.$$

Notice that both these equations amount to the same thing, namely, $p = 2^{r+1} - 1$. It looks as if whenever we can find a prime that is one less than a power of 2, say $p = 2^k - 1$, then $2^{k-1}p$ will be a perfect number. (Here k is the $r + 1$ of the displayed equations.) This is true, but we will write out a formal proof just to make sure.

Theorem 3.5. *If k is any integer such that $2^k - 1$ is prime, then $2^{k-1}(2^k - 1)$ is perfect.*

Proof. Since $2^k - 1$ is an odd prime, it is relatively prime to 2^{k-1}. Thus by the multiplicativity of σ, we have

$$\sigma(2^{k-1}(2^k - 1)) = \sigma(2^{k-1})\sigma(2^k - 1).$$

Now by Theorem 3.2

$$\sigma(2^{k-1}) = \frac{2^k - 1}{2 - 1} = 2^k - 1,$$

while since $2^k - 1$ is prime $\sigma(2^k - 1) = 2^k - 1 + 1 = 2^k$.

We see that if $n = 2^{k-1}(2^k - 1)$, then

$$\sigma(n) = (2^k - 1)2^k = 2n.$$

□

Finding More Perfect Numbers

We now have a scheme for finding perfect numbers, being able to construct one whenever we find a prime that is one less than a power of 2.

k	$2^k - 1$	prime?	perfect number
1	1	no	
2	3	yes	$2 \cdot 3 = 6$
3	7	yes	$2^2 \cdot 7 = 28$
4	15	no	
5	31	yes	$2^4 \cdot 31 = 496$
6	63	no	

The reader should check that 496 really is perfect. Clearly the determination of when $2^k - 1$ is prime is an important question in the study of perfect numbers. We defer this until the next section, however, and turn instead to the question of whether or not we are on the track of all perfect numbers.

Are All Perfect Numbers $2^{k-1}(2^k - 1)$?

That each prime of the form $2^k - 1$ gives a perfect number was known to Euclid, and the theorem we just proved appears in his *Elements*. This left the question "Are there any perfect numbers not of Euclid's form?" It took about 2000 years until progress was made on this question, and even then a complete answer was not given.

Theorem 3.6 (Euler). *Every even perfect number is of the form*

$$2^{k-1}(2^k - 1),$$

where $2^k - 1$ *is prime.*

Proof. Suppose n is an even perfect number. Then we can write

$$n = 2^r q, \quad r > 0,$$

where q is some odd integer. Using the multiplicativity of σ and the fact that q is odd, we have

$$\sigma(n) = \sigma(2^r)\sigma(q) = (2^{r+1} - 1)\sigma(q).$$

Note that we know nothing about q other than that it is odd, so we cannot evaluate $\sigma(q)$. From the fact that n is perfect, however, we have

$$\sigma(n) = 2 \cdot 2^r q = 2^{r+1} q.$$

Combining this with the previous equation and dividing through by $2^{r+1}\sigma(q)$ gives

$$\frac{2^{r+1} - 1}{2^{r+1}} = \frac{q}{\sigma(q)}.$$

This equation provides the key to the proof of this theorem. Notice that the fraction on the left is close to 1, since the numerator is only one less than the denominator. On the other hand, we would expect the fraction on the right to be smaller, since its denominator $\sigma(q)$ is a sum including q, 1, and whatever other divisors q has. In fact, we will show that q must be prime, so that it has no more divisors.

Of course, just because two fractions are equal does not mean that the same goes for their numerators and denominators; the fractions might not be in lowest terms. In the last equation, the fraction on the left is in lowest terms, since the numerator is odd, but it is not clear that the fraction on the right is. Let us suppose

$$(q, \sigma(q)) = d.$$

Then

$$q = d(2^{r+1} - 1) = d \cdot 2^{r+1} - d,$$

and

$$\sigma(q) = d \cdot 2^{r+1} = q + d.$$

Now $2^{r+1} - 1 > 1$, so d is a divisor of q other than q itself. If $d > 1$ then by definition $\sigma(q)$ is at least $q + d + 1$, which contradicts the last equation. We conclude that $d = 1$.

Setting $d = 1$ in the last equation gives $\sigma(q) = q + 1$. This clearly implies that q is prime. The previous equation then says that

$$q = 2^{r+1} - 1.$$

Setting $k = r + 1$ now yields the statement of the theorem. □

Odd Perfect Numbers

Euler's theorem only covers even perfect numbers; are there any odd ones? Nobody knows. No one has ever found an odd perfect number, but neither has anyone ever shown that none exist, and there might even be infinitely many of them. In 1991 R. B. Brent, G. L. Cohen, and H. J. J. te Riele published a paper showing that there is no odd perfect number less than 10^{300}.

Problems for Section 3.3.

1. Find the next perfect number after 496.
2. Extend the table given in this section to run from $k = 7$ to 10.
3. An early author stated that if $k > 1$ is odd then $2^k - 1$ is prime. Show that this is false.
4. An early author stated that if k is prime then $2^k - 1$ is prime. Show that this is false.

In the next five problems, (a) give the prime power factorization of $2^k - 1$; (b) list all numbers $2^d - 1$, where d divides k, $1 < d < k$.

5. $k = 4$ 6. $k = 6$ 7. $k = 8$ 8. $k = 9$ 9. $k = 10$

10. For which numbers k is $2^k + 1$ prime, $k = 1, 2, \ldots, 10$?

In the next six problems, define r by $2^r \parallel k$ if k is even. If k is odd let $r = 0$. For each k given, (a) write the prime power factorization of $2^k + 1$; (b) compute r; (c) compute $2^{2^r} + 1$.

11. $k = 3$ 12. $k = 5$ 13. $k = 6$
14. $k = 7$ 15. $k = 9$ 16. $k = 10$

B

17. Where does the proof of Euler's theorem fail if n is not even?
18. The Greeks knew of only five perfect numbers, the largest of which was 33,550,336. Of course, this is of the Euclid type, $2^{k-1}p$. What are k and p?
19. Show that if $k > 0$ then 2^k, $2^k + 1$, and $2^k - 1$ all have the same number of digits.
20. It has been shown that $2^k - 1$ is prime for $k = 11{,}213$. How many digits does this number have? (**Hint:** $\log_{10} 2 \approx 0.301029996$.)
21. Show by induction on n that if m and n are positive integers then $2^m - 1$ divides $2^{mn} - 1$.
22. Use the formula for the sum of a geometric progression from Section 1.6 to show that

$$1 + 2^m + 2^{2m} + \cdots + 2^{(n-1)m} = \frac{2^{mn} - 1}{2^m - 1}.$$

23. Show by use of congruence properties that

$$2^{mn} \equiv 1 \pmod{2^m - 1} \quad \text{for all positive integers } m \text{ and } n.$$

C

24. Show that any of the last three problems implies that for $2^k - 1$ to be prime, k must be prime.
25. Recall the function $s_{-1}(n)$ defined in the problems for Section 3.1. Show that n is perfect if and only if $s_{-1}(n) = 2$.
26. Show that if n is an odd perfect number, then $n = pa^2$, where p is prime. (**Hint:** Consider the equation $\sigma(n) = 2n$. When is $\sigma(p^k)$ odd?)
27. We could replace addition by multiplication in the definition of perfect numbers and ask when n is the product of its positive divisors other than itself. Show that this happens when $n = p^3$, p prime, or $n = pq$, p and q distinct primes.
28. Show that if $n > 1$, then n is the product of its positive divisors other than itself only in the cases mentioned in the previous problem.
29. According to Book IX of Euclid's *Elements* "If as many numbers as we please, beginning from unity, be set out continuously in double proportion until the sum of all becomes prime, and if the sum is multiplied by the last, the product will be perfect." Explain this.
30. Show that no perfect number is a square.
31. Show that if p is as in problem 26, then $p \equiv 1 \pmod 4$.
32. Show that if $\lambda > 1$, then there exists a composite n such that $\sigma(n)/n < \lambda$.

3.4 Mersenne and Fermat Numbers

DEFINITION. Mersenne numbers

If k is a positive integer we call the number $2^k - 1$ a **Mersenne number**, and denote it by M_k.

Which Mersenne Numbers Are Prime?

Of course, we are interested in this question because we have found that there is a one-to-one correspondence between Mersenne primes and even perfect numbers. Let us carry the table of the last section further.

k	$M_k = 2^k - 1$	prime?
1	1	no
2*	3	yes
3*	7	yes
4	$15 = 3 \cdot 5$	no
5*	31	yes
6	$63 = 9 \cdot 7$	no
7*	127	yes
8	$255 = 3 \cdot 5 \cdot 7$	no
9	$511 = 7 \cdot 73$	no
10	$1023 = 3 \cdot 11 \cdot 31$	no

The values of k making M_k prime have been starred in the table. They are $k = 2$, 3, 5, and 7—exactly the primes. The natural conjecture would be that M_k is prime exactly when k is prime. This is soon punctured, however, since

$$M_{11} = 2047 = 23 \cdot 89.$$

(In spite of this, Christian Wolf (1679–1754) stated in print that 2047 was prime. He even claimed that $M_9 = 511$ was prime!)

Half of the conjecture is true. For M_k to be prime, k must be prime. In fact, an examination of our table seems to indicate that if $d \mid k$, then $M_d \mid M_k$. For example,

$$2 \mid 4, \quad \text{and} \quad M_2 = 3 \mid 15 = M_4.$$

Likewise,

$$2 \mid 6 \text{ and } 3 \mid 6, \quad \text{and} \quad M_2 = 3 \text{ and } M_3 = 7 \text{ both divide } 63 = M_6.$$

Although there are many ways of proving that this works in general (see the problems at the end of the last section), by far the easiest is by means of congruences.

Theorem 3.7. *Suppose d and k are positive integers such that $d \mid k$. Then $2^d - 1 \mid 2^k - 1$. Thus if M_k is prime, k must be prime.*

RESPIGHI

ANTICHE DANZE ED ARIE

PER LIUTO

Libera trascrizione per orchestra

II Suite (sec. XVI e XVII)

I Caroso F. (1527 c. - dopo il 1605) ***Laura soave: balletto con gagliarda, saltarello e canario***

II Besardo G. B. (1567 c. - 1625 c.) ***Danza rustica***

III Autore incerto (1600..) - Mersenne M. (1588-1648) ***Campanae parisienses - Aria***

IV Gianoncelli B. (detto il Bernardello) ***Bergamasca*** **(1650)**

Figure 3.1 A tune by Mersenne appears in Respighi's *Ancient Airs and Dances*

Proof. We wish to show $2^k - 1 \equiv 0 \pmod{2^d - 1}$, or

$$2^k \equiv 1 \pmod{2^d - 1}.$$

Let $k = de$. Then since $2^d \equiv 1 \pmod{2^d - 1}$, we have

$$2^k - 1 = 2^{de} - 1 = (2^d)^e - 1 \equiv 1^e - 1 = 0 \pmod{2^d - 1},$$

where the congruence follows from Theorem 1.21. □

The History of Mersenne Primes

The Mersenne numbers are named after Marin Mersenne, a French priest who studied them and published a list (containing a number of mistakes) of primes among them. He was also a composer, and one of his melodies is used in Ottorino Respighi's *Ancient Airs and Dances.* (It is the middle part of the third movement of the Second Suite, sandwiched between music by an anonymous composer. See Figure 3.1.)

Over the years more and more Mersenne primes have been found; for example, Euler proved that M_{31} is prime. It is an unfortunate characteristic of powers

MERSENNE AND FERMAT NUMBERS

RAPHAEL M. ROBINSON

1. Mersenne numbers. The Mersenne numbers are of the form 2^n-1. As a result of the computation described below, it can now be stated that the first seventeen primes of this form correspond to the following values of n:

2, 3, 5, 7, 13, 17, 19, 31, 61, 89, 107, 127, 521, 607, 1279, 2203, 2281.

The first seventeen even perfect numbers are therefore obtained by substituting these values of n in the expression $2^{n-1}(2^n-1)$. The first twelve of the Mersenne primes have been known since 1914; the twelfth, $2^{127}-1$, was indeed found by Lucas as early as 1876, and for the next seventy-five years was the largest known prime.

Figure 3.2 Raphael Robinson tells about five new Mersenne primes in the *Proceedings of the American Mathematical Society*, 1954. He had never programmed a computer before.

of 2 that they get big quite fast; M_{31} is already more than two billion. The introduction of electronic computers has made it possible to check much larger numbers than before. As late as 1948 the largest k for which M_k was known to be prime was 127, which had been found by the French mathematician Lucas in 1876. It gave the prime

$$M_{127} = 170{,}141{,}183{,}460{,}469{,}231{,}731{,}687{,}303{,}715{,}884{,}105{,}727.$$

Then computers turned up the temperature. On January 30, 1952, during their infancy, Raphael M. Robinson applied a theoretical test originally devised by Lucas using the SWAC (the National Bureau of **S**tandards' **W**estern **A**utomatic **C**omputer) computer in Los Angeles and found two new Mersenne primes *during the first day.* He found three more that year, on June 25, October 7, and October 9. (See Figure 3.2.) This brought to 17 the number of known Mersenne primes. The SWAC computer took about a minute to determine that the smallest of the new primes, $2^{521}-1$, was indeed prime, and about an hour for the largest, $2^{2281}-1$. The total memory of the computer for both program and data was 256 words of 37 binary digits each, and this restricted checking 2^n-1 for primeness to $n < 2304$. In modern terms, the SWAC had about 1K of memory, much less than today's programmable calculators.

These finds were surpassed as computers got bigger and faster. The postage-meter stamp shown in Figure 3.3 was used by the University of Illinois to honor the discovery of the prime M_{11213} there in 1963.

In 1978 two 18-year-olds from Hayward, California—Laura Nickel and Curt Noll—used 440 hours of computer time to find the 6533-digit prime number M_{21701}. The story in Figure 3.4 appeared in *The Times* of London.

Figure 3.3 The University of Illinois used this postage meter in 1963

Prime Number Record Broken

Hayward, California, Nov. 16.—Two 18-year-old American students have discovered with the help of a computer at California State University the biggest known prime number, the number two to the 21,701st power.

Laura Nickel and Curt Noll received congratulations from Dr. Bryant Tuckerman, an American who discovered the previous record-holder among prime numbers: two to the 19,937th power.—Agence France-Presse.

Figure 3.4 Two 18-year-olds found M_{21701}

Now finding new Mersenne primes has become a group activity. In 1996 George Woltman, an Orlando, Florida, programmer, started the Great Internet Mersenne Prime Search (GIMPS), by writing and distributing software that has enabled thousands of people to participate using their personal computers. The largest prime now known is $2^{6972593} - 1$, a number of 2,098,960 digits. It was discovered June 30, 1999, by GIMPS participant Nayan Hajratwala, of Plymouth, Michigan. Hajratwala, who works for PriceWaterhouseCoopers, used a 350 MHz Pentium II IBM Aptiva computer part-time for 111 days to identify the prime.

In spite of the many Mersenne primes that have been found, no one has ever proved that there are infinitely many of them.

Summary of What is Known and Not Known about Perfect Numbers

1. If $2^k - 1$ is prime, then $2^{k-1}(2^k - 1)$ is a perfect number.

2. If n is an even perfect number, then it is of the form given in (1).

3. It is not known whether there are infinitely many even perfect numbers.

4. It is not known whether there are any odd perfect numbers at all.

5. If $2^k - 1$ is prime, then k is prime, but the converse is not true.

Fermat Numbers

Since the numbers $2^k - 1$ proved interesting, we turn to the numbers $2^k + 1$.

k	2^k+1	prime?
1*	3	yes
2*	5	yes
3	$9 = 3 \cdot 3$	no
4*	17	yes
5	$33 = 3 \cdot 11$	no
6	$65 = 5 \cdot 13$	no
7	$129 = 3 \cdot 43$	no
8*	257	yes
9	$513 = 3 \cdot 3 \cdot 3 \cdot 19$	no
10	$1025 = 5 \cdot 5 \cdot 41$	no

The values of k giving primes have again been starred in the table. So far, they are exactly the powers of 2: 1, 2, 4, and 8. Looking at the composite numbers in the second column we see the same divisors recurring, namely, 3 and 5; and these are themselves of the form $2^k + 1$. If k is odd, 3 seems to divide 2^{k+1}, while if k is even (but not divisible by 4), 5 is a divisor.

Our table suggests the conjecture that if $k = ab$, where a is odd, then

$$2^b + 1 \mid 2^k + 1.$$

Although this may be proved by direct division, a congruence proof is much easier, and provides another demonstration of the usefulness of the congruence notation.

Theorem 3.8. *If k, a, and b are positive integers such that $k = ab$, where a is odd, then $2^b + 1 \mid 2^k + 1$. In particular, if $2^k + 1$ is prime, then k is 0 or a power of 2.*

Proof. We wish to show $2^k + 1 \equiv 0 \pmod{2^b + 1}$, or

$$2^k \equiv -1 \pmod{2^b + 1}.$$

Now $2^b \equiv -1 \pmod{2^b + 1}$, so

$$2^k = 2^{ab} = (2^b)^a \equiv (-1)^a \equiv -1 \pmod{2^b + 1},$$

where the first congruence follows from Theorem 1.21 and the second from the fact that a is odd.

To prove the last sentence of the theorem we note that if $k > 0$ is not a power of 2 then we can take $a > 1$. Thus

$$1 < 2^b + 1 < 2^k + 1,$$

and so $2^k + 1$ has a positive divisor other than 1 and itself. □

DEFINITION. Fermat numbers

If r is a nonnegative integer we call

$$2^{2^r} + 1$$

a **Fermat number**, and denote it by F_r.

Examples. $F_0 = 2^{2^0} + 1 = 3$, $F_1 = 2^2 + 1 = 5$, $F_2 = 2^4 + 1 = 17$, and $F_3 = 2^8 + 1 = 257$. ♢

Are All Numbers $2^{2^r} + 1$ Prime?

Fermat conjectured that they are. Since 2 raised to a power of 2 is involved, the Fermat numbers get large much faster than even the Mersenne numbers, and so it is much more difficult to tell whether they are prime or not. The number $F_4 = 65{,}537$ is not too big, and can be shown to be a prime, but $F_5 = 2^{32} + 1$ is already greater than 4 billion.

It took about 100 years after Fermat's conjecture for Euler to succeed in showing that F_5 was composite. He used a theoretical method rather than merely checking the 10-digit number for factors.

We will give a congruence proof that 641 divides F_5 involving very little arithmetic. First notice that

$$641 = 640 + 1 = 2^7 5 + 1$$

and

$$641 = 625 + 16 = 5^4 + 2^4.$$

Thus $2^7 5 \equiv -1 \pmod{641}$ and $2^4 \equiv -5^4 \pmod{641}$. Then (all congruences being modulo 641 and using Theorems 1.16 and 1.21),

$$F_5 = 2^{32} + 1 = 2^{28} 2^4 + 1 \equiv (2^7)^4(-5^4) + 1 = -(2^7 5)^4 + 1 \equiv -(-1)^4 + 1 = 0.$$

We see Fermat's conjecture was incorrect. Worse than that, no other primes F_r have as yet been found! Many other values of r are known for which F_r is composite, but aside from these numerical results not much is known. It is possible that F_r is prime only for $r = 0$, 1, 2, 3, and 4; but it is also possible that other Fermat primes exist, even infinitely many.

Just as Mersenne primes apply to the study of perfect numbers, Fermat primes arise in other parts of mathematics. They are connected with the construction of regular polygons in geometry. For example, the reason that a regular pentagon can be constructed with straightedge and compass but a regular 7-sided polygon cannot is that 5 is a Fermat prime but 7 is not. In fact the great German mathematician Gauss proved that a regular polygon with $n > 2$ sides is constructible if and only if n is a power of 2 times a product (possibly empty) of distinct Fermat primes.

Problems for Section 3.4.

A

In the first five problems, find the prime power factorization of $2^k + 1$.

1. $k = 11$ 2. $k = 12$ 3. $k = 13$ 4. $k = 14$ 5. $k = 15$

The next eight problems are based on the material of Sections 3.3 and 3.4. For each statement, write T (true), F (false), or N (nobody knows). Give counterexamples for the false statements.

6. If p is prime, then $2^p - 1$ is prime.
7. If $2^k(2^{k+1} - 1)$ is perfect, then $2^k - 1$ is prime.
8. There are only four Fermat primes.
9. There are infinitely many prime Fermat numbers.
10. Every perfect number is of the form $2^{k-1}(2^k - 1)$.
11. If $2^k + 1$ is prime, then k is a power of 2.
12. There are infinitely many odd perfect numbers.
13. If $d \mid k$, then $M_d \mid M_k$.

B

Let a, b, and k be positive integers, a odd, and suppose $k = ab$. The next two problems give alternate proofs that $2^b + 1 \mid 2^k + 1$, by proving the formula $2^k + 1 = (2^b + 1)(1 - 2^b + 2^{2b} - \cdots - 2^{b(a-2)} + 2^{b(a-1)})$.

14. Prove the formula by applying the formula for the sum of a geometric progression to the factor on the right.
15. Prove the formula by multiplying out the right side.

16. Let $a > 1$. Show that if d and e are positive integers, then $a^d - 1$ divides $a^{de} - 1$. (**Hint:** Show $a^{de} \equiv 1 \pmod{a^d - 1}$.)
17. Prove that if $a > 0$, $k > 1$, and $a^k - 1$ is prime, then $a = 2$ and k is prime.
18. Prove that if $a > 1$, $e > 0$, and $q > 0$, q odd, then $a^e + 1$ divides $a^{qe} + 1$. (**Hint:** Show $a^{qe} \equiv -1 \pmod{a^e + 1}$.)
19. Show that if $a > 1$, $k > 0$, and $a^k + 1$ is prime, then a is even and k is a power of 2.

C

20. Show that if $0 \le j < k$, then F_j divides $F_k - 2$.
21. Show that if $0 \le j < k$, then F_j and F_k are relatively prime.
22. Use the last problem to show there exist infinitely many primes.
23. Prove by induction on n that $F_n = F_0 F_1 \cdots F_{n-1} + 2$ for all positive integers n.
24. Show that if $n > 1$, then $\sigma(n)/n < \prod p/(p-1)$, where the product is over all primes p dividing n.
25. Show that if n is an odd perfect number, then at least 3 primes divide n.

3.5 The Euler Phi Function

Measuring "Primeness"

There has hardly been anything we have done up to now in which the prime numbers have not played an important role. Although being prime is an all-or-nothing proposition, we have the feeling that some numbers are more composite than others. For example, $60 = 2 \cdot 2 \cdot 3 \cdot 5$ seems further from primeness than $62 = 2 \cdot 31$. One way we might quantify this feeling is with $\tau(n)$, the number of positive divisors of n. Only if n is prime can $\tau(n)$ be 2; for composite numbers it is bigger. In a certain sense, the bigger $\tau(n)$ is, the further n is from being prime. For example, $\tau(62) = 4$, while $\tau(60) = 12$.

Another way to measure the same thing depends on the fact that if p is prime, then $(p, n) > 1$ if and only if $p \mid n$. This property characterizes the primes, for if a is composite, it has a divisor n such that $1 < n < a$. Then $(a, n) > 1$ but a doesn't divide n.

Other things being equal, we expect $(a, n) = 1$ to be a more common occurrence when a is prime than when a is composite. As an example let us write out the first few n such that $(7, n) = 1$. The easiest way is to start writing out all the integers, crossing out the multiples of 7.

1	2	3	4	5	6	~~7~~
8	9	10	11	12	13	~~14~~
15	16	17	18	19	20	~~21~~

Let us now do the same thing with 6 instead of 7. Since $(6, n) > 1$ if and only if some prime divides both n and $6 = 2 \cdot 3$, it suffices to cross out the multiples of 2 and 3.

1	~~2~~	~~3~~	~~4~~	5	~~6~~
7	~~8~~	~~9~~	~~10~~	11	~~12~~
13	~~14~~	~~15~~	~~16~~	17	~~18~~

The n such that $(6, n) = 1$ appear to be much rarer.

The reader has probably noticed a pattern in the two arrays just presented. Either all or none of the numbers in each column have been crossed out. This is no surprise if we recall that Theorem 1.20 says that if $b \equiv b' \pmod{a}$, then $(a, b) = (a, b')$. In particular $(a, b) = 1$ if and only if $(a, b') = 1$.

In the first array, we listed the numbers in 7 columns, so those in any given column were all congruent modulo 7. In the second array, the columns consisted of numbers congruent modulo 6. By the argument in the previous paragraph, either all the numbers in a given column are relatively prime to a ($= 6$ or 7), or else none are.

Since we cannot count all the n such that $(a, n) = 1$, there being infinitely many of them, let us merely count those in the "first row"; after that the pattern repeats anyway.

DEFINITION. Euler ϕ function

Suppose a is a positive integer. We define $\phi(a)$ to be the number of integers n, $1 \le n \le a$, such that $(a,n) = 1$. The function ϕ is called the **Euler phi function**.

Examples. By what we have already seen, $\phi(6) = 2$ and $\phi(7) = 6$. Clearly $\phi(p) = p - 1$ for any prime p. By definition $\phi(1) = 1$.

Let us compute $\phi(12)$. Since $12 = 2^2 \cdot 3$, it suffices to cross out the multiples of 2 and 3 from among the first 12 integers:

$$1 \quad \not{2} \quad \not{3} \quad \not{4} \quad 5 \quad \not{6} \quad 7 \quad \not{8} \quad \not{9} \quad \not{10} \quad 11 \quad \not{12}$$

We see $\phi(12) = 4$. ◇

A Shortcut to Computing $\phi(n)$

Computing $\phi(12)$ by our "crossing out" method amounted to eliminating the same numbers as in computing $\phi(6)$, the multiples of 2 and 3. Thus the numbers left are exactly the same as in the first two rows of our table for 6 above, and the pattern from 7 to 12 repeats that of 1 through 6.

If we had noticed this at the start we could have cut our work in half. We could have crossed out the multiples of 2 and 3 among the first 6 integers, leaving 2 numbers, then doubled this count of 2 to get $\phi(12) = 4$. In other words, $\phi(12) = 2\phi(6)$.

In the same way, $\phi(18) = 3\phi(6) = 6$. This works since 18 has no prime divisor that does not already divide 6. (Note that $\phi(30) = 8 \ne 5\phi(6)$.)

Proposition G. *If k and a are positive integers such that all primes dividing k also divide a, then $\phi(ka) = k\phi(a)$.*

Proof. Consider the array

$$\begin{array}{cccc} 1 & 2 & \cdots & a \\ a+1 & a+2 & \cdots & 2a \\ & & \vdots & \\ (k-1)a+1 & (k-1)a+2 & \cdots & ka. \end{array}$$

Let us cross out the entries that are not relatively prime to ka in order to compute $\phi(ka)$. This amounts to crossing out everything not relatively prime to a, since $(a,n) > 1$ if and only if some prime divides both a and n, and the same primes divide ka as divide a. By Theorem 1.20 the pattern of crossings-out is the same in each row. Since there are $\phi(a)$ entries left in the first row, there must be $k \cdot \phi(a)$ entries left in all. Thus $\phi(ka) = k \cdot \phi(a)$. □

Example. Since 3 is prime, $\phi(3) = 3-1 = 2$. Thus $\phi(27) = \phi(9\cdot 3) = 9\cdot\phi(3) = 9 \cdot 2 = 18$. We can apply Proposition G because 3 is the only prime dividing 9. ◇

A Formula for ϕ

Proposition G goes a long way toward a formula for $\phi(n)$. For example, suppose we want to compute $\phi(1{,}000{,}000)$. We have

$$\phi(10^6) = \phi(2^6 5^6) = \phi(2^5 5^5 2 \cdot 5) = 2^5 5^5 \cdot \phi(10),$$

but we still must compute $\phi(10)$ by actual count. The computation of $\phi(n)$ can always be reduced in the same way to that of a product of distinct primes. If we could pull the p_1 out of $\phi(p_1 p_2 \dots p_t)$ just as we learned to pull the k out of $\phi(ka)$ under the hypothesis of Proposition G, we could write a complete formula.

Let us see if we can modify the proof of Proposition G to compute $\phi(pa)$, where p is a prime *not* dividing a. As before, let us write out the integers from 1 to pa.

$$\begin{array}{cccc} 1 & 2 & \cdots & a \\ a+1 & a+2 & \cdots & 2a \\ & & \vdots & \\ (p-1)a+1 & (p-1)a+2 & \cdots & pa \end{array}$$

Again let us imagine crossing out the n such that $(pa, n) > 1$. The difference here is that $(pa, n) > 1$ is not equivalent to $(a, n) > 1$; the latter implies the former but not conversely. Thus of the $p \cdot \phi(a)$ integers left after crossing out all n such that $(a, n) > 1$, more still must be eliminated, namely, the multiples of p. These are just $1p$, $2p$, ... , ap. Of these, those kp such that $(k, a) > 1$ have already been eliminated, and there are just $\phi(a)$ more to cross out in order that $\phi(pa)$ be properly counted. Thus

$$\phi(pa) = p \cdot \phi(a) - \phi(a) = (p-1)\phi(a).$$

We have proved the following theorem:

Proposition H. *If a is a positive integer, p is prime, and p doesn't divide a, then*

$$\phi(pa) = (p-1)\phi(a).$$

Examples.

$$\begin{aligned} \phi(10^6) &= 10^5\phi(2 \cdot 5) = 10^5(2-1)\phi(5) = 40{,}000 \\ \phi(60) &= \phi(2^2 3 \cdot 5) = 2 \cdot \phi(2 \cdot 3 \cdot 5) \\ &= 2(2-1)\phi(3 \cdot 5) = 2(3-1)\phi(5) = 16 \\ \phi(62) &= \phi(2 \cdot 31) = (2-1)\phi(31) = 30 \end{aligned}$$

◇

Theorem 3.9. *If $n = p_1^{k_1} p_2^{k_2} \cdots p_t^{k_t}$, where $p_1, p_2, \ldots, p_t$ are distinct primes and $k_1, k_2, \ldots, k_t$ are positive integers, then*

$$\phi(n) = \prod_{i=1}^{t} (p_i - 1) p_i^{k_i - 1}.$$

Proof. Let $Q = p_2^{k_2} p_3^{k_3} \cdots p_t^{k_t}$. Then, using Propositions G and H, we have

$$\phi(n) = \phi(p_1^{k_1} Q) = p_1^{k_1 - 1} \phi(p_1 Q) = p_1^{k_1 - 1}(p_1 - 1)\phi(Q).$$

The same technique may be applied to the prime powers in Q until we arrive at the formula of the theorem. □

Examples.

$$\begin{aligned} \phi(60) &= \phi(2^2 3 \cdot 5) = (2-1)2(3-1)(5-1) = 16, \\ \phi(62) &= \phi(2 \cdot 31) = (2-1)(31-1) = 30, \\ \phi(360) &= \phi(2^3 3^2 5) = (2-1)2^2(3-1)3(5-1) = 96. \end{aligned}$$

◇

The Function Phi Is Multiplicative

It is not hard to prove that ϕ is multiplicative now that we have a formula for it; in fact, we leave this proof for the exercises. Note that this reverses our practice of late of first proving a function multiplicative, and then using this fact to derive a formula.

Theorem 3.10. *The function ϕ is multiplicative.*

Now consider the function F defined by

$$F(n) = \sum_{d|n} \phi(d).$$

Note that if p is prime and k is a positive integer, then

$$\begin{aligned} F(p^k) = \sum_{d|p^k} \phi(d) = \phi(1) + \phi(p) + \cdots + \phi(p^k) \\ &= 1 + (p-1) + (p-1)p + \cdots + (p-1)p^{k-1} \\ &= 1 + (p-1)(1 + p + p^2 + \cdots + p^{k-1}) \\ &= 1 + (p-1)\sigma(p^{k-1}) = 1 + (p-1)\frac{p^k - 1}{p - 1} = p^k. \end{aligned}$$

Since F is multiplicative by the previous theorem and Theorem 3.3, we have proved the following result.

Theorem 3.11. *If n is any positive integer, then*

$$\sum_{d|n} \phi(d) = n.$$

Problems for Section 3.5.

A

For each n given in the first eight problems, write out the integers from 1 to n, then cross out those not relatively prime to n. Use this to compute $\phi(n)$.

1. $n = 4$
2. $n = 8$
3. $n = 16$
4. $n = 12$
5. $n = 3$
6. $n = 9$
7. $n = 15$
8. $n = 10$

Which of the next six statements are true by Proposition G?

9. $\phi(3 \cdot 12) = 3\phi(12)$
10. $\phi(4 \cdot 12) = 4\phi(12)$
11. $\phi(8 \cdot 12) = 8\phi(12)$
12. $\phi(12 \cdot 8) = 12\phi(8)$
13. $\phi(5 \cdot 12) = 5\phi(12)$
14. $\phi(5 \cdot 12) = 4\phi(12)$

Which of the next six statements are true by Proposition H?

15. $\phi(3 \cdot 12) = 3\phi(12)$
16. $\phi(3 \cdot 12) = 2\phi(12)$
17. $\phi(5 \cdot 12) = 5\phi(12)$
18. $\phi(5 \cdot 12) = 4\phi(12)$
19. $\phi(25 \cdot 12) = 24\phi(12)$
20. $\phi(35 \cdot 12) = 24\phi(12)$

In the next eight problems, compute $\phi(n)$ for each n by using the formula.

21. $n = 100$
22. $n = 200$
23. $n = 27$
24. $n = 210$
25. $n = 999$
26. $n = 127$
27. $n = 91$
28. $n = 1001$

29. This problem illustrates the proof of Proposition H, with $p = 5$ and $a = 6$. Write out 1, 2, ... , 30 in rows of 6.

    (a) Cross out all n such that $(6, n) > 1$. How many integers are left?

    (b) Circle the multiples of 5. How many are there?

    (c) How many multiples of 5 are not already crossed out?

    (d) If the multiples of 5 are crossed out how many of the 30 numbers are not crossed out?

30. Compute $\phi(n)$ for $n = 100$, 200, 27, and 210, using the formula of problem 34 below.
31. Confirm Theorem 3.11 for $n = 12$.
32. Confirm Theorem 3.11 for $n = 30$.

B

33. Use Theorem 3.9 to prove that ϕ is multiplicative.
34. Show that for all n

$$\phi(n) = n \prod_{\substack{p|n \\ p \text{ prime}}} \left(1 - \frac{1}{p}\right).$$

In the next four problems, find all solutions x to $\phi(x) = n$ and prove that you have found them all. (It has been conjectured that there is no n such that $\phi(x) = n$ has exactly one solution x, but this has never been proved.)

35. $n = 1$ 36. $n = 2$ 37. $n = 4$ 38. $n = 14$

39. Show that the formula of Theorem 3.9 fails if some of the ks are allowed to be 0.
40. Let $\varepsilon > 0$. Prove there exists $n > 1$ such that

$$1 - \frac{\phi(n)}{n} < \varepsilon.$$

41. Prove directly from the definition of ϕ that if p is prime, then $\phi(p^k) = (p-1)p^{k-1}$.
42. What are all n such that $\phi(n)$ is odd?
43. What are all n such that $\phi(2n) = \phi(n)$?
44. What are all n such that $\phi(2n) = \phi(3n)$?
45. Show that if $\phi(n)$ divides $n - 1$, then the square of no prime divides n.
46. Show that $\phi(n^2) = n\phi(n)$ for all positive integers n.
47. Show that if 4 doesn't divide $\phi(n)$, then n is 1, 2, 4, or of the form p^k or $2p^k$, where p is a prime congruent to 3 (mod 4).
48. Prove that if $m \mid n$, then $\phi(m) \mid \phi(n)$.
49. Prove that $\tau(n) + \phi(n) \leq n + 1$ for all positive integers n.
50. Show that if n is prime, then $\phi(n) + \sigma(n) = n\tau(n)$.
51. C Show that if $\phi(n) + \sigma(n) = n\tau(n)$, then n is prime.
52. Show that for all positive integers n,

$$\sum_{d \mid n} M(d)\phi(d) = \prod_{\substack{p \mid n \\ p \text{ prime}}} (2 - p),$$

where M is as in the problems for Section 3.2.

53. Show that $\phi'' = \sigma$, in the notation of the problems for Section 3.2.
54. Show that if n is fixed, then $\phi(x) = n$ has only a finite number of solutions.
55. Find N such that $n \geq N$ implies $\phi(n) \geq 100$.
56. Show that $\phi(n) \to \infty$ as $n \to \infty$.
57. Show that if n is such that $\phi(x) = n$ has exactly one solution x, then 36 divides x.
58. Another proof that there are infinitely many primes uses the function ϕ. Assume there are only finitely many primes, and let P be their product. Derive a contradiction by computing $\phi(P)$.

In the next five problems, let $N(n)$, $E(n)$, and $O(n)$ be, respectively, the number of solutions, number of even solutions, and number of odd solutions x to $\phi(x) = 2^n$.

59. Confirm that $N(n) = n + 2$ for $n = 0, 1, 2, 3, 4, 5$. (This formula holds for $n < 32$.)
60. Show that if $n \geq 1$, then $N(n) = E(n-1) + 2 \cdot O(n)$.
61. Show that $N(n) = n + 2$ for all $n \geq 0$ if and only if $O(n) = 1$ for all $n \geq 0$.

62. Show the odd number x is a solution of $\phi(x) = 2^n$ for some $n > 0$ if and only if x is the product of distinct Fermat primes.
63. Show that $N(32) = 32$.

3.6 The Möbius Inversion Formula

In Section 3.1 we saw that finding a formula for a numerical function was easier if we knew that it was multiplicative, for in that case we had only to determine its value at prime powers. Theorem 3.3 states that if two numerical functions f and F are related by

$$F(n) = \sum_{d|n} f(d), \quad n = 1, 2, \ldots, \tag{3.4}$$

then F is multiplicative whenever f is. If f is a simple function, then it may be easy to show it is multiplicative, from which the multiplicativity of F follows. Examples are $f(n) = 1$, $F(n) = \tau(n)$ and $f(n) = n$, $F(n) = \sigma(n)$.

Given a numerical function F, can we always find a function f so that (3.4) holds? We will answer this question below.

Finding f from F

Let us start with as simple an example as we can. Consider the function

$$F(n) = 1, \quad n = 1, 2, \ldots.$$

We will try to find a function f such that (3.4) holds. Taking $n = 1$ in (3.4) we get $F(1) = f(1)$, and so $f(1) = F(1) = 1$. Taking $n = 2$ yields $F(2) = f(2) + f(1)$, or $1 = f(2) + 1$, and so we must have $f(2) = 0$. Likewise using $n = 3$, we get $F(3) = f(1) + f(3)$, or $1 = 1 + f(3)$, and so $f(3) = 0$ also. In the same way, $F(4) = f(1) + f(2) + f(4)$, or $1 = 1 + 0 + f(4)$, implying that $f(4) = 0$.

It appears that we may have $f(1) = 1$ and $f(n) = 0$ for $n > 1$. Let us not lose sight of our logical position. We have *assumed* the existence of a function f satisfying (3.4) when $F(n)$ is identically 1. Our calculations show that if such a function f exists, then of necessity $f(1) = 1$, and $f(2) = f(3) = f(4) = 0$. Such computations do not establish the existence of such a function, however.

An analogous situation is the way many people would "solve" the equation

$$3x + 2 = -4$$

by writing

$$\begin{aligned} 3x &= -4 - 2 = -6, \\ x &= -\frac{6}{3} = -2. \end{aligned}$$

This sequence of computations shows only that if $3x + 2 = -4$ has a solution, then it must be $x = -2$. After all, what is the "x" that is manipulated in the equations? It must be a number assumed to satisfy the original equation. We must substitute -2 for x in the original equation to show that -2 actually is a solution.

The problem at hand may be handled in the same way. Let us define the function f by

$$f(n) = \begin{cases} 1, & n = 1, \\ 0, & n > 1. \end{cases}$$

Then it is almost obvious that

$$\sum_{d|n} f(d) = 1 \quad \text{for all positive integers n,}$$

and so f is a solution to our problem.

Of course, our calculation of $f(n)$ for $n = 1$, 2, 3, and 4 was not worthless; it suggested to us what function to try for f. It also indicated that if an f satisfying (3.4) exists, then its values are determined by those of F, and so are unique. This is true for any function F, and in fact was proved back in Section 3.2. (did you remember?) as Theorem 3.4.

The Möbius Function

Let us try another example of finding f from F to see if we can somehow generalize the process. Suppose $F(1) = 1$, while if $n > 1$, then $F(n) = 0$. (Thus we are taking F to be the function called f in the previous example.) Let us assume there exists a function f satisfying (3.4). We will try to find a formula for f.

It is easy to show directly that the function F is multiplicative. We might hope that the function f we seek is also multiplicative. In any case, let us try to see how such an f would behave when applied to a power of a prime.

As usual, $f(1) = F(1) = 1$. Let p be prime. Then from (3.4) $F(p) = f(1) + f(p)$, or $0 = 1 + f(p)$, from which we see that $f(p) = -1$. Likewise $F(p^2) = f(1) + f(p) + f(p^2)$, or $0 = 1 - 1 + f(p^2)$, and so $f(p^2) = 0$. In the same way, $F(p^3) = f(1) + f(p) + f(p^2) + f(p^3)$, or $0 = 1 - 1 + 0 + f(p^3)$, so $f(p^3) = 0$ also. Continuing in this way, we see that $f(p^k) = 0$ for all integers $k > 1$.

Now suppose f were multiplicative. Let $n = p_1^{r_1} p_2^{r_2} \cdots p_t^{r_t}$, the ps distinct primes and the rs positive integers. We would have $f(n) = f(p_1^{r_1}) \cdots f(p_t^{r_t})$. If any exponent r_i is greater than 1, then $f(p_i^{r_i}) = 0$, and so $f(n) = 0$ also. On the other hand, if $r_1 = r_2 = \cdots = r_t = 1$, then $f(n) = (-1)(-1) \cdots (-1) = (-1)^t$. The function f we have found is important in number theory and is conventionally denoted by μ. It takes its name from A. F. Möbius, a student of Gauss.

DEFINITION. Möbius function

The **Möbius function** is the function μ defined by

$$\mu(n) = \begin{cases} 1 & \text{if } n = 1, \\ 0 & \text{if } p^2 \mid n \text{ for some prime } p, \\ (-1)^t & \text{if } n = p_1 p_2 \cdots p_t, \text{ the } p\text{s distinct primes.} \end{cases}$$

For example, $\mu(28) = 0$ since $2^2 \mid 28$, $\mu(29) = (-1)^1 = -1$, and $\mu(30) = (-1)^3 = -1$.

Since in our last example it was assumed that a function f existed satisfying (3.4) for $F(1) = 1$ and $F(n) = 0$ for $n > 1$, we must check that μ really works.

Theorem 3.12. *The function μ is multiplicative, and*

$$\sum_{d|n} \mu(d) \tag{3.5}$$

is 1 if $n = 1$ and 0 if $n > 1$.

Proof. First we show μ is multiplicative. Suppose $(a, b) = 1$. We wish to show that $\mu(ab) = \mu(a)\mu(b)$. If either a or b is 1, this is easy. Likewise if the square of some prime divides a or b, then both sides are 0. The only remaining case is when $a = p_1 p_2 \cdots p_t$ and $b = q_1 q_2 \cdots q_u$, the ps and qs being distinct primes. Then no p is a q since $(a, b) = 1$. Thus $\mu(ab) = \mu(p_1 \cdots p_t q_1 \cdots q_u) = (-1)^{t+u} = (-1)^t(-1)^u = \mu(a)\mu(b)$.

Now let $G(n)$ stand for the summation of (3.5). Since we have proved that μ is multiplicative, so is G by Theorem 3.3. But by definition, $G(1) = \mu(1) = 1$, while if p is prime and $k > 0$

$$G(p^k) = \mu(1) + \mu(p) + \mu(p^2) + \cdots + \mu(p^k) = 1 - 1 + 0 + \cdots + 0 = 0,$$

and so since G is multiplicative $G(n) = 0$ for $n > 1$. □

The Möbius Inversion Formula

Now we consider the general problem of finding a function f satisfying (3.4) for all positive integers n, given a numerical function F. Taking $n = 1$ in (3.4) tells us that we must have $f(1) = F(1)$. Now let p be prime. Then by (3.4) with $n = p$ we have $F(p) = f(1) + f(p)$, or $F(p) = F(1) + f(p)$, and so

$$f(p) = F(p) - F(1). \tag{3.6}$$

Likewise taking $n = p^2$ yields

$$F(p^2) = f(1) + f(p) + f(p^2) = F(1) + (F(p) - F(1)) + f(p^2),$$

and so

$$f(p^2) = F(p^2) - F(p). \tag{3.7}$$

The reader should make a similar computation to check that we must have

$$f(p^3) = F(p^3) - F(p^2). \tag{3.8}$$

Note that equations (3.6), (3.7), and (3.8) can all be expressed as

$$f(n) = F(n) - F(n/p) \quad \text{for } n = p,\ p^2, \text{ or } p^3, \tag{3.9}$$

and we might suspect that this formula holds for n any power of a prime.

Now let us see what happens when more than one prime is involved, say $n = pq$, where p and q are distinct primes. From (3.4) and (3.6) we see

$$\begin{aligned} F(pq) &= f(1) + f(p) + f(q) + f(pq) \\ &= F(1) + (F(p) - F(1)) + (F(q) - F(1)) + f(pq), \end{aligned}$$

and so $f(pq) = F(pq) - F(p) - F(q) + F(1)$, which we could write

$$f(n) = F(n) - F(n/p) - F(n/q) + F(n/pq) \quad \text{for } n = pq. \tag{3.10}$$

The reader should check that (3.10) also holds for $n = p^2q$, and also that if p, q, and r are distinct primes and $n = pqr$, then

$$\begin{aligned} f(n) = F(n) - F(n/p) - F(n/q) - F(n/r) \\ + F(n/pq) + F(n/pr) + F(n/qr) - F(n/pqr). \end{aligned} \tag{3.11}$$

It appears we can express $f(n)$ in the form $\sum m(d)F(n/d)$, where d runs through the positive divisors of n, and m is a function that takes on only the values 0, 1, and -1. In fact, it appears that $m(d) = 0$ unless d is 1 or a product of distinct primes, and in the latter case $m(d)$ is -1 or 1 based on whether an odd or even number of primes is involved. But this exactly decribes the Möbius function.

Theorem 3.13 (the Möbius inversion formula). *Let F be any numerical function. Define f by*

$$f(n) = \sum_{d|n} \mu(d) F\left(\frac{n}{d}\right), \quad n = 1, 2, 3, \ldots .$$

Then

$$F(n) = \sum_{d|n} f(d) \quad \textit{for all integers } n.$$

Proof. To prove this theorem we must substitute the function f defined by the first equation into the right side of the second one and confirm that we actually get $F(n)$ for all positive integers n. This substitution gives

$$\sum_{d|n} \left(\sum_{e|d} \mu(e) F\left(\frac{d}{e}\right) \right).$$

Notice that since $d \mid n$ the number d/e is always some divisor k of n. We will regroup this sum as

$$\sum_{k|n} h(k) F(k)$$

by collecting the terms containing $F(k)$ for a fixed divisor k of n. For example, if $n = 4$, we could write

$$\begin{aligned}
\sum_{d|4}\left(\sum_{e|d}\mu(e)F\left(\frac{d}{e}\right)\right) &= \sum_{e|1}\mu(e)F\left(\frac{1}{e}\right)+\sum_{e|2}\mu(e)F\left(\frac{2}{e}\right)+\sum_{e|4}\mu(e)F\left(\frac{4}{e}\right)\\
&= \mu(1)F(1)+\mu(1)F(2)+\mu(2)F(1)+\mu(1)F(4)+\mu(2)F(2)+\mu(4)F(1)\\
&= (\mu(1)+\mu(2)+\mu(4))F(1)+(\mu(1)+\mu(2))F(2)+\mu(1)F(4)\\
&= (1-1+0)F(1)+(1-1)F(2)+1F(4)=F(4).
\end{aligned}$$

In general, the coefficient $h(k)$ of $F(k)$ will be $\sum\mu(e)$, where the sum runs over all pairs d, e such that $d \mid n$, $e \mid d$, and $d/e = k$. Of course, if d/e is the integer k, then it is automatic that $e \mid d$, and so we can forget this condition. Also once we choose e, the equation $d/e = k$ completely determines d; we need only figure out which values of e make $d = ek$ a divisor of n. But $ek \mid n$ if and only if n/ek is an integer, or e is a divisor of n/k. Thus

$$\sum_{d|n}\left(\sum_{e|d}\mu(d)F\left(\frac{n}{d}\right)\right)=\sum_{k|n}h(k)F(k)=\sum_{k|n}\left(\sum_{e|n/k}\mu(e)\right)F(k).$$

Now by the previous theorem the inner sum is 1 when $n/k = 1$, and 0 otherwise. Thus the last expression is simply $F(n)$, which is what we wanted to prove. □

Note that the Möbius inversion formula does not depend on whether or not the function F is multiplicative.

Example. Find a formula for the numerical function f such that

$$\sum_{d|n} f(d) = n$$

for all positive integers n.

By the Möbius inversion formula, we have

$$f(n)=\sum_{d|n}\mu(d)\frac{n}{d}.$$

Let $n = p_1^{k_1}p_2^{k_2}\dots p_t^{k_t}$, where the ps are distinct primes. Suppose d is a positive divisor of n. We have $\mu(d) = 0$ except when d is a product of distinct primes, and we can associate each subset S of $\{1, 2, \dots, t\}$ with such a $d = \prod_{i\in S} p_i$. Denote the number of elements of S by $|S|$. Then

$$\begin{aligned}
f(n) &= \sum_{S\subseteq\{1,2,\dots,t\}}\mu\left(\prod_{i\in S}p_i\right)\frac{n}{\prod_{i\in S}p_i}=n\sum_{S\subseteq\{1,2,\dots,t\}}\frac{(-1)^{|S|}}{\prod_{i\in S}p_i}\\
&= n(1-\frac{1}{p_1}-\dots-\frac{1}{p_t}+\frac{1}{p_1p_2}+\cdots)=n(1-\frac{1}{p_1})(1-\frac{1}{p_2})\dots(1-\frac{1}{p_t})\\
&= \prod_{i=1}^{t}p_i^{k_i}\prod_{i=1}^{t}(1-\frac{1}{p_i})=\prod_{i=1}^{t}(p_i-1)p_i^{k_i-1}.
\end{aligned}$$

But we recognize this as the formula for the Euler phi function, as given by Theorem 3.9. Indeed, the last calculation brings old news, since we proved in Theorem 3.11 in the previous section that $\sum_{d|n} \phi(d) = n$ for all positive integers n. ♢

Problems for Section 3.6.

A

1. Compute $\mu(n)$ for $n = 1, 2, \ldots, 12$.
2. Compute $\mu(n)$ for $n = 13, 14, \ldots, 24$.
3. Find all n, $25 < n < 50$, such that $\mu(n) = 1$.
4. Find all n, $50 < n < 75$, such that $\mu(n) = 0$.
5. Find all nonprimes $n < 100$ with $\mu(n) = -1$.
6. Find all $n < 100$ such that $\mu(n)\mu(n+1)\mu(n+2) = 1$.
7. Confirm that the summation (3.5) of this section is 0 for $n = 12$.
8. Confirm that the summation (3.5) of this section is 0 for $n = 30$.

B

In the next four problems, a function F is given. Find a formula for the function f satisfying equation (3.4) of this section.

9. $F(n) = \mu(n)$.
10. $F(n) = (\mu(n))^2$.
11. $F(n) = 1$ if n is odd and 0 if n is even.
12. $F(n) = n^2$.

The next two problems should be done without assuming any results of this section.

13. Suppose $F(n) = 1$ for all positive integers n, and that f is a function satisfying (3.4) of this section. Prove by induction on k that if p is prime, then $f(p^k) = 0$ for all positive integers k.
14. Suppose $F(1) = 1$ and $F(n) = 0$ for $n > 1$ and that f is a function satisfying (3.4) of this section. Prove by induction on k that if p is prime, then $f(p^k) = 0$ for all integers $k > 1$.
15. Show that if $n = p_1^{k_1} \cdots p_t^{k_t}$, the ps distinct primes and the ks positive integers, then the formula (3.5) of this section has exactly 2^t nonzero terms.
16. Show that if F and f are functions satisfying (3.4), then

$$f(n) = \sum_{d|n} \mu\left(\frac{n}{d}\right) F(d) \quad \text{for all positive integers } n.$$

C

17. Show that if n is any positive integer, then $\mu(n)\mu(n+1)\mu(n+2)\mu(n+3) = 0$.
18. Prove that there are infinitely many integers n such that $\mu(n)+\mu(n+1) = 0$.
19. Prove that there are infinitely many integers n such that $\mu(n)+\mu(n+1)+\mu(n+2) = 0$.

20. If $n = p_1^{k_1} \cdots p_t^{k_t}$, the ps distinct primes and the ks positive integers, define $F(n)$ to be $(-1)^t$. Find the function f satisfying (3.4) of this section.

21. Define $F(n)$ to be 1 if $n = m^2$ for some integer m and 0 otherwise. Find the function f satisfying (3.4) of this section.

22. Prove that there exist infinitely many n such that $\mu(n) + \mu(n+1) = -1$.

In the following problems, $\lfloor X \rfloor$ denotes the greatest integer $\leq X$, called the ***integer part*** *or* ***floor*** *of X, where X is any real number.*

23. Show that if X is a positive real number and p and q are distinct primes, the the number of positive integers $\leq X$ not divisible by p or q is $\lfloor X \rfloor - \lfloor X/p \rfloor - \lfloor X/q \rfloor + \lfloor X/pq \rfloor$.

24. Let $p_1, p_2, \ldots, p_t$ be distinct primes and X a positive real number. Show that the number of positive integers $\leq X$ not divisible by any of $p_1, \ldots, p_t$ is

$$\lfloor X \rfloor - \lfloor \frac{X}{p_1} \rfloor - \lfloor \frac{X}{p_2} \rfloor - \cdots - \lfloor \frac{X}{p_t} \rfloor$$
$$+ \lfloor \frac{X}{p_1 p_2} \rfloor + \lfloor \frac{X}{p_1 p_3} \rfloor + \cdots + \lfloor \frac{X}{p_{t-1} p_t} \rfloor - \lfloor \frac{X}{p_1 p_2 p_3} \rfloor - \cdots + (-1)^t \lfloor \frac{X}{p_1 p_2 \cdots p_t} \rfloor.$$

25. Let $n = p_1^{k_1} \cdots p_t^{k_t}$, the ps distinct primes, and the ks positive integers. Show that the number of positive integers $\leq X$ relatively prime to n is $\sum \mu(d) \lfloor X/d \rfloor$, where in the summation d runs through the positive divisors of n.

26. Show that the sum in the previous problem has at most 2^t nonzero terms.

27. Let $\pi(X)$ denote the number of primes $\leq X$. Show that if p_i is the ith prime (so $p_1 = 2$, $p_2 = 3$, etc.), if $p_t < X$, and if $n = p_1 p_2 \cdots p_t$, then $\pi(X) \leq t + \sum \mu(d) \lfloor X/d \rfloor$, where in the summation d runs through the positive divisors of n.

28. Show that with the hypotheses of the previous problem

$$\pi(X) \leq t + 2^t + X \left(1 - \frac{1}{p_1}\right) \left(1 - \frac{1}{p_2}\right) \cdots \left(1 - \frac{1}{p_t}\right).$$

29. Show that $\pi(X)/X \to 0$ as $X \to \infty$. (**Hint:** See problem 28 in Section 2.5.)

30. Show that if f is multiplicative and m and n are positive integers, then $f(m)f(n) = f((m,n))f([m,n])$.

Carl Friedrich Gauss

1777–1855

If a poll were taken among mathematicians as to who was the all-time greatest practitioner of the subject, Gauss would probably get more votes than anyone else. The son of a German bricklayer, Gauss conceived most of his discoveries before the age of 20, but spent the rest of his life polishing and refining them. His motto, "few, but ripe," reflected his habit of retaining, even concealing, his results until he thought them sufficiently perfected (or until he felt the world was ready for them). For example, Gauss laid the foundations of non-Euclidean geometry, but hid his researches until others published the idea.

Gauss considered number theory to be the most interesting part of mathematics, but spent much of his energies on more applied subjects. He invented the method of least squares independently from Legendre, and, in his position as director of the astronomical observatory at Göttengen, calculated the orbits of heavenly bodies. He also worked in electricity and magnetism, inventing with Wilhelm Weber a working telegraph in 1833.

Early in his life, Gauss had to choose between a career in mathematics or the study of languages. His discovery, at the age of 19, of the constructability of the regular 17-gon, led him to choose the former. This was the first advance along this line since the time of Euclid, more than 2000 years earlier.

Gauss was also the first to put complex numbers on a logical footing, and he introduced to analysis what today is considered mathematical rigor.

Gauss's life was long and generally quiet and happy. He disliked travel, but retained his interest in literature and world affairs, reading extensively in English and Russian. His accomplishments, both in and out of mathematics, were so so extensive that many must be omitted in a short account such as this one.

CARL FRIEDRICH GAUSS

Chapter 4

The Algebra of Congruence Classes

4.1 Solving Linear Congruences

What Does It Mean to Solve a Congruence?

The reader knows what it means to solve an equation. To "solve" the equation

$$x^2 + x - 2 = 0,$$

for example, we must find all values of x making the equation true. We will consider the similar problem where the equality sign is replaced by "$\equiv$." In general, we will restrict ourselves to congruences of the form

$$F(x) \equiv 0 \pmod{b},$$

where $F(x)$ is a polynomial in x with integer coefficients.

Suppose x_1 satisfies the congruence

$$x^2 + x - 2 \equiv 0 \pmod{7},$$

and suppose that $x_2 \equiv x_1 \pmod{7}$. Then by Theorem 2.3

$$x_2^2 + x_2 - 2 \equiv x_1^2 + x_1 - 2 \equiv 0 \pmod{7},$$

and so x_2 is also a solution of the congruence. Thus if we find one solution x, we immediately have infinitely many more solutions, namely, all integers congruent to x modulo 7. Since we cannot write all of these down, we will content ourselves with identifying any one of them, and the one we identify will serve as a representative for the whole class of solutions congruent modulo 7.

Since Theorem 2.3 applies to any polynomial with integer coefficients, we have the following general result.

Theorem 4.1. *Let x_1, x_2, and $b > 0$ be integers with $x_1 \equiv x_2 \pmod{b}$, and let F be a polynomial with integer coefficients. Then x_1 is a solution to*

$$F(x) \equiv 0 \pmod{b}$$

if and only if x_2 is.

It is convenient at this point to introduce names for various sets of integers of importance for a given modulus.

DEFINITION. congruence class, complete residue system

Suppose $b > 0$. By a **congruence class modulo** b we mean the set of all integers congruent modulo b to some fixed integer k. By a **complete residue system modulo** b we mean a set of integers such that exactly one integer in the set is in each congruence class modulo b.

Examples. There are three congruence classes modulo $b = 3$, namely,

$$\begin{aligned} &\{\ldots, -6, -3, 0, 3, 6, 9, \ldots\}, \\ &\{\ldots, -5, -2, 1, 4, 7, 10, \ldots\}, \quad \text{and} \\ &\{\ldots, -4, -1, 2, 5, 8, 11, \ldots\}. \end{aligned}$$

One complete residue system modulo 3 is $\{0, 1, 2\}$. Another is $\{1, 2, 3\}$. Another is $\{10, -1, 6\}$. ◇

If we write out the integers in a table with rows of length b, as has proved useful several times, then the columns are exactly the congruence classes modulo b. Taking $b = 4$, for example, we get

$$\begin{array}{rrrr} . & . & . & . \\ . & -6 & -5 & -4 \\ -3 & -2 & -1 & 0 \\ 1 & 2 & 3 & 4 \\ 5 & 6 & 7 & 8 \\ 9 & 10 & 11 & . \\ . & . & . & ., \end{array}$$

and each column is a congruence class modulo 4. If we choose exactly one number from each column, say 5, -2, 11, and 0, we get a complete residue system modulo 4.

It is easy to see that there are exactly b congruence classes modulo b, and that any complete residue system modulo b also consists of exactly b elements.

Counting Solutions to a Congruence

Recall that if F is a polynomial with integral coefficients, then any integer congruent modulo b to a solution to

$$F(x) \equiv 0 \pmod{b} \tag{4.1}$$

is also automatically a solution, and so doesn't deserve separate attention. This is the reason for the following definition.

DEFINITION. complete solution, least complete solution

We call $x_1, x_2, \dots, x_k$ a **complete solution** to the congruence (4.1) in case it is the set of all solutions in some complete residue system modulo b. It is the **least complete solution** in case it is the set of all solutions among 0, 1, ... , $b-1$.

Examples. Consider the congruence

$$4x - 18 \equiv 0 \pmod{6}.$$

Among the integers 1, 2, 3, 4, 5, 6 exactly 3 and 6 satisfy the congruence. Thus $x = 3, 6$ is a complete solution. Another complete solution is $x = -3, 0$, since these are all solutions in the complete residue system −3, −2, −1, 0, 1, 2. Another is $x = 9, 72$. The least complete solution is $x = 0, 3$. ♢

Note that it is possible to match up the elements of any two complete solutions so that corresponding numbers are congruent modulo b. For example,

$$9 \equiv 3 \quad \text{and} \quad 72 \equiv 6 \pmod{6}.$$

On the other hand, $x = -3, 9$ is not a complete solution to $4x - 18 \equiv 0 \pmod{6}$ since −3 and 9 are in the same congruence class.

The Congruence $ax \equiv c \pmod{b}$

In this chapter we will consider only congruences $F(x) \equiv 0 \pmod{b}$, where F is a linear polynomial. Such a congruence can be put in the form $ax \equiv c \pmod{b}$. To solve this congruence we must look for values of x such that

$$b \mid c - ax,$$

that is, such that

$$c - ax = by$$

for some integer y. This is the equation $ax + by = c$ that we analyzed completely in Section 1.4. There we decided that

1. The equation $ax + by = c$ has a solution if and only if (a, b) divides c.

2. A solution can be found explicitly by using the Euclidean algorithm to find (a, b) and then solving the equations backward.

3. If x_0, y_0 is a particular solution, then all solutions are given by $x = x_0 + bt/d$, $y = y_0 - at/d$, where t runs through the integers and $d = (a, b)$.

We will illustrate the use of (1) and (2) to find a solution to

$$147x \equiv 77 \pmod{161}.$$

Here $a = 147$, $b = 161$, and $c = 77$. First we use the Euclidean algorithm to find $(147, 161)$.

$$\begin{aligned} 161 &= 1 \cdot 147 + 14 \\ 147 &= 10 \cdot 14 + 7 \\ 14 &= 2 \cdot 7 + 0 \end{aligned}$$

We see that $(147, 161) = 7$, and since 7 divides 77, the congruence has a solution.

Now we solve the equations backward as follows:

$$\begin{aligned} 7 &= 147 - 10 \cdot 14 \\ &= 147 - 10(161 - 147) \\ &= 11 \cdot 147 - 10 \cdot 161. \end{aligned}$$

Multiplying through by $77/7 = 11$ gives

$$77 = 121 \cdot 147 - 110 \cdot 161.$$

We have found one solution to the congruence, namely, $x = 121$.

Now let us see how (3) tells us how to find a complete solution to the congruence

$$ax \equiv c \pmod{b}.$$

We know that if x_0 is one solution to the corresponding equation, then all solutions are given by

$$x = x_0 + \frac{bt}{d},$$

where $d = (a, b)$ and t runs through the integers. This gives infinitely many values of x, but we only want to find the solutions in some complete residue system modulo b; that is, we want at most one solution in any congruence class modulo b.

How is it possible for us to have

$$x_0 + \frac{bt}{d} \equiv x_0 + \frac{bt'}{d} \pmod{b}$$

for integers t and t'? This amounts to the fact that b divides

$$x_0 + \frac{bt'}{d} - \left(x_0 + \frac{bt}{d}\right) = \frac{(t' - t)b}{d}.$$

Then $(t' - t)b/d = kb$, or $t' - t = kd$ for some integer k; that is,

$$t' \equiv t \pmod{d}.$$

Thus using only values of t incongruent modulo d will certainly give us solutions x incongruent modulo b. Furthermore, we will not miss any solutions this way, for it can be shown that

$$t' \equiv t \pmod{d}$$

implies that

$$x_0 + \frac{bt}{d} \equiv x_0 + \frac{bt'}{d} \pmod{b};$$

see the exercises at the end of this section.

We see that if the values of t are restricted to lie in some complete residue system modulo $d = (a, b)$, then the values of $x = x_0 + bt/d$ will be incongruent modulo b, and will comprise a complete solution.

Theorem 4.2. *Consider the congruence*

$$ax \equiv c \pmod{b}.$$

This has a solution if and only if (a, b) divides c. If x_0 is any solution, then a complete solution is given by the numbers

$$x = x_0 + \frac{bt}{(a, b)},$$

where t runs through any complete residue system modulo (a, b).

Example. We return to the congruence

$$147x \equiv 77 \pmod{161},$$

for which we have already found a particular solution $x = 121$. We also found that $(a, b) = (147, 161) = 7$. Then a complete solution to the congruence is given by

$$x = 121 + \frac{161t}{7} = 121 + 23t,$$

where t runs through any complete residue system modulo 7. Letting $t = 0, 1, 2, 3, 4, 5, 6$ yields the complete solution

$$x = 121, 144, 167, 190, 213, 236, 259.$$

(Note that taking $t = 7$ gives $x = 282$; but $282 \equiv 121 \pmod{161}$.) The least complete solution is $x = 121$, 144, $167 - 161 = 6$, $190 - 161 = 29$, $213 - 161 = 52$, $236 - 161 = 75$, $259 - 161 = 98$. ◇

Example. Suppose we are asked for a complete solution to

$$15x \equiv 50 \pmod{100}.$$

By inspection we find the particular solution $x_0 = 10$. We also note that $(15, 100) = 5$. Then a complete solution is given by

$$x = 10 + \frac{100t}{5} = 10 + 20t,$$

where $t = 0,\ 1,\ 2,\ 3,\ 4$. We get

$$x = 10, 30, 50, 70, 90$$

as a complete solution. (It is also the least complete solution.) $\diamondsuit$

Example. Find a complete solution to the congruence

$$1485x \equiv 999 \pmod{2222}.$$

Here

$$\begin{aligned} 2222 &= 1 \cdot 1485 + 737 \\ 1485 &= 2 \cdot 737 + 11 \\ 737 &= 67 \cdot 11 + 0, \end{aligned}$$

so $(1485, 2222) = 11$. But 11 doesn't divide 999 and so there are no solutions. $\diamondsuit$

Example. Find a complete solution to

$$45x \equiv 3 \pmod{48},$$

where the solutions are to be as small as possible in absolute value.

We note by inspection the solution $x_0 = -1$. Since $(45, 48) = 3$, a complete solution is

$$x = -1 + \frac{48t}{3} = -1 + 16t,$$

where $t = 0,\ 1,\ 2$. This gives

$$x = -1, 15, 31.$$

Now -1 and 15 are the elements smallest in absolute value in their congruence classes, but to fulfill the conditions of the problem we must replace 31 by $31 - 48 = -17$. Thus the answer is

$$x = -17, -1, 15.$$

Problems for Section 4.1.

A

In the first eight problems, tell whether or not each sequence given is a complete residue system modulo 6.

1. $-3, -2, -1, 1, 2, 3$
2. $0, 3, 6, 9, 12, 15$
3. $0, 5, 10, 15, 20, 25$
4. $0, 1, 2, 3, 4, 5, 6$
5. $1, 2, 4, 8, 3, 5$
6. $3, 8, 7, 18, 23, 22$
7. $7, 14, 21, 28, 35$
8. $5, -5, 4, -4, 3, 60$

In the next eight problems, tell whether each list of values of x forms a complete solution to the given congruence.

9. $x = 4;\ 2x \equiv 8 \pmod 6$
10. $x = 4, 10;\ 2x \equiv 8 \pmod 6$
11. $x = 4;\ 2x \equiv 8 \pmod 9$
12. $x = 4, 10;\ 10x \equiv 4 \pmod{12}$
13. $x = 0;\ 6x \equiv 4 \pmod{12}$
14. $x = 3, 7;\ 3x \equiv 9 \pmod{12}$
15. $x = 3;\ x^2 \equiv 2 \pmod 7$
16. $x = 7;\ x^2 + x \equiv 1 \pmod 5$

In the next eight problems, tell the number of elements in a complete solution.

17. $5x \equiv 100 \pmod{55}$
18. $11x \equiv 14 \pmod{23}$
19. $91x \equiv 169 \pmod{143}$
20. $91x \equiv 169 \pmod{140}$
21. $1001x \equiv 143 \pmod{99}$
22. $48x \equiv 128 \pmod{1000}$
23. $x^2 + x + 1 \equiv 0 \pmod{14}$
24. $x^2 + x + 1 \equiv 0 \pmod{91}$

In the next 10 problems, give the least complete solution to the congruence.

25. $6x \equiv 2 \pmod 8$
26. $36x \equiv 30 \pmod{42}$
27. $25x \equiv 100 \pmod{35}$
28. $143x \equiv 169 \pmod{110}$
29. $27x \equiv -18 \pmod{15}$
30. $51x \equiv 0 \pmod{17}$
31. $3x \equiv 18 \pmod{18}$
32. $253x \equiv 341 \pmod{299}$
33. $165x \equiv 84 \pmod{221}$
34. $441x \equiv 465 \pmod{640}$

35. Find all x, $0 \leq x \leq 9$, such that $5x \equiv 15 \pmod{10}$.
36. Find all x, $|x| < 5$, such that $5x \equiv 20 \pmod{10}$.
37. Find all x, $100 \leq x < 110$, such that $6x \equiv 2 \pmod{10}$.
38. Find all x, $100 \leq x < 110$, such that $6x \equiv 9 \pmod{15}$.

B

In the next three problems, you may want to invoke the **pigeon-hole principle***: if $n+1$ objects are put into n boxes, then some box must contain more than one object. Assume $b > 0$.*

39. Show that if S and T are two congruence classes modulo b, then either $S = T$ or else S and T have no elements in common.
40. Show that if S has b elements and no two elements of S are congruent modulo b, then S is a complete residue system modulo b.

41. Show that if S has b elements and if S contains an element in each congruence class modulo b, then S is a complete residue system modulo b.

True-False. In the next seven problems, state whether each statement is true or false. Give counterexamples for the false statements. Assume that a, b, and c are positive, $(a, b) = 1$, and that $x_1, x_2, \ldots, x_b$ is a complete residue system modulo b.

42. The list $x_1 + c, x_2 + c, \ldots, x_b + c$ is a complete residue system modulo b.
43. The list $cx_1, cx_2, \ldots, cx_b$ is a complete residue system modulo b.
44. The list $ax_1, ax_2, \ldots, ax_b$ is a complete residue system modulo b.
45. The list $-x_1, -x_2, \ldots, -x_b$ is a complete residue system modulo b.
46. There exists a complete residue system modulo b, say $y_1, y_2, \ldots, y_b$, such that $|y_i| \leq b/2$, $i = 1, 2, \ldots, b$.
47. There exists a complete residue system modulo b, say $y_1, y_2, \ldots, y_b$, such that $|y_i| < b/2$, $i = 1, 2, \ldots, b$.
48. The list $x_1^2, x_2^2, \ldots, x_b^2$ is a complete residue system modulo b.

C

49. Show that if $(a, b) = 1$ and if $x_1, x_2, \ldots, x_b$ is a complete residue system modulo b, then so is $ax_1, ax_2, \ldots, ax_b$.
50. Prove or disprove: If P is a polynomial such that $P(x)$ is an integer for all integers x, and if $y \equiv y' \pmod{b}$, then $P(y) \equiv P(y') \pmod{b}$.
51. Prove that if d and b are positive integers such that $d \mid b$ and $t' \equiv t \pmod{d}$, then $x_0 + bt/d \equiv x_0 + bt'/d \pmod{b}$.
52. Suppose m objects are put into n boxes. Consider the following statements: (a) each box has at least one object; (b) no box has more than one object; (c) $m \geq n$; (d) $m \leq n$. Which pairs of statements imply the other two statements?

4.2 The Chinese Remainder Theorem

Simultaneous Congruences

Suppose a and b are relatively prime positive integers. By Theorem 4.2 we can choose x so that ax is in any congruence class we want modulo b. For that matter we can also choose y so that by is in any congruence class we want modulo a. To emphasize the symmetry between a and b we will change the notation and consider relatively prime positive integers b_1 and b_2.

By Theorem 4.2 we can find x_1 such that

$$b_2 x_1 \equiv 1 \pmod{b_1}.$$

Then

$$b_2 x_1 c_1 \equiv c_1 \pmod{b_1},$$

where c_1 is completely arbitrary. Likewise we can find x_2 such that

$$b_1 x_2 \equiv 1 \pmod{b_2},$$

and then

$$b_1 x_2 c_2 \equiv c_2 \pmod{b_2},$$

where c_2 is arbitrary.

The important implication of the above congruences is that we can make the sum $z = b_2 x_1 c_1 + b_1 x_2 c_2$ simultaneously congruent to whatever we want modulo both b_1 and b_2. For

$$z = b_2 x_1 c_1 + b_1 x_2 c_2 \equiv b_2 x_1 c_1 + 0 \equiv c_1 \pmod{b_1},$$

and

$$z = b_2 x_1 c_1 + b_1 x_2 c_2 \equiv 0 + b_1 x_2 c_2 \equiv c_2 \pmod{b_2}.$$

The ability to find an integer z such that

$$z \equiv c_1 \pmod{b_1} \quad \text{and} \quad z \equiv c_2 \pmod{b_2},$$

where c_1 and c_2 are arbitrary, has many applications.

Example. Professor Snabley feeds his pet python every four days and bathes it once a week. This week he fed it on Tuesday and washed it on Wednesday. When, if ever, will he feed and wash the python on the same day? How often will this happen?

Let us number the days, with this Tuesday as day 1. Then the snake will be fed on days 1, 5, 9, ... , and, in general, on day z exactly when

$$z \equiv 1 \pmod{4}.$$

Since he washes the snake every seven days beginning with Wednesday (day 2), day z is a washday exactly when

$$z \equiv 2 \pmod{7}.$$

Because the moduli 4 and 7 are relatively prime, the simultaneous congruences

$$z \equiv 1 \pmod{4} \quad \text{and} \quad z \equiv 2 \pmod{7} \tag{4.2}$$

will have a solution by the preceding analysis. Let us take $b_1 = 4$, $c_1 = 1$, $b_2 = 7$, and $c_2 = 2$. We start by finding an integer x_1 such that $b_2 x_1 \equiv 1 \pmod{b_1}$, or $7x_1 \equiv 1 \pmod{4}$. One solution is $x_1 = 3$. Likewise we want x_2 such that $b_1 x_2 \equiv 1 \pmod{b_2}$, or $4x_2 \equiv 1 \pmod{7}$. A solution is $x_2 = 2$.

Now according the computations at the beginning of this section a solution to (4.2) is

$$z = b_2 x_1 c_1 + b_1 x_2 c_2 = 7 \cdot 3 \cdot 1 + 4 \cdot 2 \cdot 2 = 37.$$

We easily check that indeed

$$37 \equiv 1 \pmod{4} \quad \text{and} \quad 37 \equiv 2 \pmod{7}.$$

Thus the snake will be washed and fed on day 37.

How often the snake gets washed and fed on the same day is another question. By (4.2) it happens on day z exactly when

$$z \equiv 1 \equiv 37 \pmod{4} \quad \text{and} \quad z \equiv 2 \equiv 37 \pmod{7}.$$

This means $4 \mid 37 - z$ and $7 \mid 37 - z$, which is clearly equivalent to $4 \cdot 7 = 28 \mid 37 - z$. Thus the snake will bathe and eat when $z \equiv 37 \pmod{28}$, every 28 days. The next time this will happen will be on day $9 = 37 - 28$, which is Wednesday of next week. ◊

The congruence property of 4 and 7 mentioned at the end of the last example works for any relatively prime positive integers. The proof of the following theorem is left for the exercises.

Theorem 4.3. *Let b_1 and b_2 be relatively prime positive integers, and let z and z' be any integers. Then $z \equiv z' \pmod{b_1}$ and $z \equiv z' \pmod{b_2}$ if and only if $z \equiv z' \pmod{b_1 b_2}$*

More Than Two Congruences

Now let us try to solve

$$z \equiv \begin{cases} c_1 \pmod{b_1} \\ c_2 \pmod{b_2} \\ c_3 \pmod{b_3}, \end{cases} \tag{4.3}$$

where $(b_1, b_2) = (b_1, b_3) = (b_2, b_3) = 1$. It seems natural to expect z to be the sum of three terms, two of which are congruent to 0 modulo b_i for any particular i. Consider

$$z = b_2b_3x_1c_1 + b_1b_3x_2c_2 + b_1b_2x_3c_3.$$

Note that whatever x_1, x_2, and x_3 are,

$$z \equiv b_2b_3x_1c_1 + 0 + 0 \equiv b_2b_3x_1c_1 \pmod{b_1}.$$

Now since b_1 is relatively prime to b_2 and b_3, we have $(b_1, b_2b_3) = 1$ by Corollary 1.12. Thus we can choose x_1 so that

$$b_2b_3x_1 \equiv 1 \pmod{b_1},$$

and so

$$b_2b_3x_1c_1 \equiv c_1 \pmod{b_1}.$$

A similar argument works for the moduli b_2 and b_3. Let us make things look simpler by introducing new notation. Let $B = b_1b_2b_3$, and set $B_1 = b_2b_3 = B/b_1$, $B_2 = B/b_2$, and $B_3 = B/b_3$. Then our solution to (4.3) is $z = B_1x_1c_1 + B_2x_2c_2 + B_3x_3c_3$, where x_i satisfies $B_ix_i \equiv 1 \pmod{b_i}$ for $i = 1, 2, 3$.

Example. Consider the system

$$\begin{aligned} z &\equiv 2 \pmod{3} \\ z &\equiv 5 \pmod{4} \\ z &\equiv -3 \pmod{7}. \end{aligned}$$

This is solvable, since $(3, 4) = (3, 7) = (4, 7) = 1$. We have

$$b_1 = 3, \quad b_2 = 4, \quad b_3 = 7, \quad \text{and} \quad c_1 = 2, \quad c_2 = 5, \quad c_3 = -3,$$

so $B = 3 \cdot 4 \cdot 7 = 84$, $B_1 = 84/3 = 28$, $B_2 = 21$, and $B_3 = 12$. The congruences

$$\begin{aligned} 28x_1 \equiv x_1 &\equiv 1 \pmod{3} \\ 21x_2 \equiv x_2 &\equiv 1 \pmod{4} \\ 12x_3 \equiv 5x_3 &\equiv 1 \pmod{7} \end{aligned}$$

have the solutions $x_1 = 1$, $x_2 = 1$, and $x_3 = 3$. Thus a solution to the original system is

$$z = B_1x_1c_1 + B_2x_2c_2 + B_3x_3c_3 = 28 \cdot 1 \cdot 2 + 21 \cdot 1 \cdot 5 + 12 \cdot 3 \cdot (-3) = 53.$$

$\diamondsuit$

DEFINITION. relatively prime in pairs

We say the integers $b_1, b_2, \ldots, b_n$ are **relatively prime in pairs** in case $(b_i, b_j) = 1$ whenever $i \neq j$. For example, 3, 4, and 7 are relatively prime in pairs, but 3, 4, and 6 are not, even though no integer greater than 1 divides all three.

The following theorem extends the last result to n congruences. It gets its name from the fact that it was introduced into European mathematics from Chinese writings. It was known to the Chinese at least as early as the first century A.D. We leave the proof for the exercises at the end of this section.

Theorem 4.4 (the Chinese remainder theorem). *Let $b_1, b_2, \dots, b_n$ be integers greater than 0, relatively prime in pairs, and let $c_1, c_2, \dots, c_n$ be any integers. Consider the system of congruences*

$$\begin{aligned} z &\equiv c_1 \pmod{b_1} \\ z &\equiv c_2 \pmod{b_2} \\ &\vdots \\ z &\equiv c_n \pmod{b_n}. \end{aligned}$$

Let $B = b_1 b_2 \cdots b_n$ and $B_i = B/b_i$ for $i = 1, 2, \dots, n$. Let x_i be a solution to

$$B_i x_i \equiv 1 \pmod{b_i}$$

for $i = 1, 2, \dots, n$. Then a solution to the original system of congruences is

$$z = \sum_{i=1}^{n} B_i x_i c_i.$$

Furthermore, z' is another solution if and only if $z' \equiv z \pmod{B}$.

Example. Find all solutions z, $0 < z < 500$, to

$$\begin{aligned} z &\equiv 1 \pmod{2} \\ z &\equiv 2 \pmod{3} \\ z &\equiv 3 \pmod{5} \\ z &\equiv 4 \pmod{7}. \end{aligned}$$

We take $b_1 = 2$, $b_2 = 3$, $b_3 = 5$, $b_3 = 7$, $c_1 = 1$, $c_2 = 2$, $c_3 = 3$, $c_4 = 4$, $B = 210$, $B_1 = 105$, $B_2 = 70$, $B_3 = 42$, and $B_4 = 30$. We must solve the congruences

$$\begin{aligned} 105x_1 &\equiv 1 \pmod{2} \\ 70x_2 &\equiv 1 \pmod{3} \\ 42x_3 &\equiv 1 \pmod{5} \\ 30x_4 &\equiv 1 \pmod{7}. \end{aligned}$$

We can take $x_1 = 1$, $x_2 = 1$, $x_3 = 3$, and $x_4 = 4$. Thus a solution is

$$z = 105 \cdot 1 \cdot 1 + 70 \cdot 1 \cdot 2 + 42 \cdot 3 \cdot 3 + 30 \cdot 4 \cdot 4 = 1103.$$

Since $B = 210$ all solutions are of the form $1103 + 210t$. Solving $0 < 1103 + 210t < 500$ leads to

$$-5\tfrac{53}{210} < t < -2\tfrac{183}{210},$$

and $t = -5, -4, -3$. The corresponding solutions are $z = 53$, 263, and 473. ◇

Reduced Residue Systems

Now we return to the consideration of a fixed modulus b. A complete residue system contains exactly one element in each congruence class modulo b, but sometimes we are only interested in integers relatively prime to b.

DEFINITION. reduced residue system

By a **reduced residue system modulo** b we mean all integers relatively prime to b in some complete residue system modulo b.

Examples. Since 1, 2, 3, 4, 5, 6 is a complete residue system modulo 6, a reduced residue system modulo 6 consists of 1 and 5. Others are $\{7,5\}$, $\{-1,1\}$, and $\{61,65\}$. One reduced residue system modulo 5 is $\{1,2,3,4\}$, another is $\{1,12,23,34\}$. ♢

Note that any two reduced residue systems modulo b have the same number of elements, namely, the number of congruence classes of integers relatively prime to b. Since the number of integers in the complete residue system $1, 2, \ldots, b$ relatively prime to b is by definition $\phi(b)$:

Any reduced residue system modulo b has $\phi(b)$ elements.

Theorem 4.5. *If a and $b > 0$ are relatively prime integers, then as x runs through a complete or reduced residue system modulo b, so does ax.*

Proof. As x runs through a complete residue system modulo b the integers ax are distinct modulo b by the cancellation theorem. Since there are b of them, they also form a complete residue system modulo b.

Likewise as x runs through a reduced residue system modulo b the integers ax are distinct modulo b by the same argument. They are all relatively prime to b by Corollary 1.12. Since there are $\phi(b)$ of them, they also form a reduced residue system modulo b. □

Example. A complete residue system modulo $b = 8$ is $\{-2,-1,0,1,2,3,4,5\}$, and a reduced residue system is $\{-1,1,3,5\}$. We take $a = 3$. The reader should check that $\{-6,-3,0,3,6,9,12,15\}$ is also a complete residue system (mod 8), and that $\{-3,3,9,15\}$ is a reduced residue system (mod 8). ♢

Problems for Section 4.2.

A

Solve the systems of the first 12 problems.

1. $z \equiv 20 \pmod 9$
   $z \equiv -3 \pmod{10}$
   $0 \le z < 90$
2. $z \equiv -2 \pmod 8$
   $z \equiv 4 \pmod{11}$
   $0 \le z < 88$
3. $x \equiv 2 \pmod 9$
   $x \equiv -2 \pmod{11}$
   $0 \le x < 150$
4. $x \equiv 8 \pmod{13}$
   $x \equiv 10 \pmod 5$
   $0 \le x < 50$

5. $x \equiv 2 \pmod 3$
$x \equiv 2 \pmod 5$
$x \equiv 2 \pmod 7$
$0 < x < 200$

6. $x \equiv 3 \pmod 4$
$x \equiv 3 \pmod 5$
$x \equiv 3 \pmod 7$
$0 < x < 200$

7. $x \equiv 1 \pmod 5$
$x \equiv -2 \pmod 7$
$x \equiv 7 \pmod 2$
$0 \le x < 70$

8. $t \equiv 2 \pmod 3$
$t \equiv 1 \pmod 4$
$t \equiv -1 \pmod 5$
$0 \le t < 60$

9. $z \equiv 3 \pmod 2$
$z \equiv 1 \pmod 3$
$z \equiv -1 \pmod 5$
$z \equiv -4 \pmod 7$
$0 \le z < 210$

10. $q \equiv 1 \pmod 3$
$q \equiv 2 \pmod 4$
$q \equiv -5 \pmod 5$
$q \equiv 5 \pmod 7$
$0 \le q < 420$

11. $q \equiv 3 \pmod 4$
$q \equiv 5 \pmod 7$
$q \equiv -3 \pmod{11}$
$0 \le q < 502$

12. $e \equiv 3 \pmod 8$
$e \equiv 6 \pmod 7$
$e \equiv 2 \pmod 5$
$0 \le e < 316$

In the next six problems, tell whether the given list is a reduced residue system modulo 5.

13. $-1, 0, 1, 2$
14. $-2, 0, 2, 4$
15. $2, 4, 6, 8$
16. $4, 6, -4, 2$
17. $7, 14, 21$
18. $2, 4, 6, 8, 12$

B

19. Solve $x \equiv 2 \pmod 8$
$x \equiv 1 \pmod 7$
$x \equiv 3 \pmod 6$
$0 \le x < 336$.

20. Solve $3x \equiv 9 \pmod 6$
$2x \equiv 1 \pmod 5$
$x \equiv 2 \pmod 7$
$0 \le x < 210$.

21. Let p, q, and r be distinct primes. Show there exists n such that $p \mid n$, $q \mid n+1$, and $r \mid n+2$.
22. Give a proof of Theorem 4.3 based on the theorems of Sections 1.1 and 1.2.
23. Show that if $(a,b) = (a,c) = (b,c) = 1$ and if $a \mid m$, $b \mid m$, and $c \mid m$, then $abc \mid m$.
24. Suppose $(a,b) = 1$. Show that as x runs through a complete residue system (mod b) and y runs through a complete residue system (mod a), then $ax + by$ runs through a complete residue system (mod ab).
25. Prove by induction on n that if $b_1, b_2, \ldots, b_n$ are integers, relatively prime in pairs, and if $b_i \mid m$ for $i = 1, 2, \ldots, n$, then $b_1 b_2 \cdots b_n \mid m$.
26. Prove the previous problem using the fundamental theorem of arithmetic.

The next four problems complete the proof of the Chinese remainder theorem. Assume the notation and assumptions of the theorem in them.

27. Show that the congruence $B_i x_i \equiv 1 \pmod{b_i}$ is solvable for $i = 1, 2, \ldots, n$.

28. Show that the number z defined in the theorem satisfies the original system of congruences.
29. Show that if z is a solution to the system and $z' \equiv z \pmod{B}$, then z' is also a solution.
30. Show that if z and z' are solutions to the system of congruences, then $z' \equiv z \pmod{B}$.

True-False. In the next six problems tell which statements are true, and give counterexamples for those that are false. Assume that $x_1, x_2, \ldots, x_t$ is a reduced residue system modulo b, and that $(a, b) = 1$.

31. The list $2x_1, 2x_2, \ldots, 2x_t$ is a reduced residue system (mod b).
32. The list $x_1 + 2, x_2 + 2, \ldots, x_t + 2$ is a reduced residue system (mod b).
33. The list $x_1 + a, x_2 + a, \ldots, x_t + a$ is a reduced residue system (mod b).
34. The list $x_1^2, x_2^2, \ldots, x_t^2$ is a reduced residue system (mod b).
35. The list $-x_1, -x_2, \ldots, -x_t$ is a reduced residue system (mod b).
36. We have $b = \phi(t)$.

37. Give a set of positive integers that is a reduced residue system both modulo 5 and modulo 8.
38. Give a reduced residue system modulo 9 consisting entirely of prime numbers.
39. Professor Crittenden buys a new car every three years; he bought his first in 1981. He gets a sabbatical leave every seven years, starting in 1992. When will he first get both during a leap year?
40. Senator McKinley was first elected in 1980. His reelection is assured unless the campaign coincides with an attack of the seven-year-itch such as hit him in 1983. When must he worry first?
41. A certain baby demands to be fed every five hours. On Tuesday, July 22, its mother watched the start of the NBC *Today* program at 7 A.M. while feeding the baby. When will this happen again?
42. A homeowner drains her hot water heater every 33 days and changes the filter in her heat pump every 56 days. She did the first on January 3, 2000, and the second on January 5, 2000. When after this does she first do both on the same day?
43. On the Fourth of July Sam took a red pill at 8 A.M. and a green pill at noon. The next day he took a white pill at 10 P.M. He takes the red, green, and white pills every 5, 7, and 24 hours, respectively. How often does he take all three together? When in August will this first happen, if at all?
44. Leroy gave each of his four children the same amount of money. One spent all but 13 cents on 5-cent candy bars. (This is a very old problem.) The second spent all but 3 cents on 6-cent iceballs. The third spent all but 2 cents on 11-cent comic books. The fourth bought a $3 game, but didn't have enough money to buy another. How much money did each child get?

45. Seven children found some jelly beans. When they divided them as evenly as possible among themselves, 1 was left over. They gave it to the cat, who ate it. Then the sister of one of the children appeared, so they redivided the jelly beans among the 8 children. This time there were 6 left over, which the cat ate. Then another child appeared. When they divided the jelly beans among the 9 children, 6 were left over. What is the least number of jelly beans they can have started with?

C

46. Suppose $(a, b) = 1$. Show that as x runs through a reduced residue system modulo b and y runs through a reduced residue system modulo a, then $ax + by$ runs through a reduced residue system modulo ab.

47. Use the previous problem to give a proof that the Euler phi-function is multiplicative.

In the next three problems, suppose that a, b, and c are positive integers, relatively prime in pairs.

48. Show that $(x, c) = (y, a) = (z, b) = 1$ implies $(abx + bcy + acz, abc) = 1$.

49. Show that as x, y, and z run through complete residue systems modulo c, a, and b, respectively, then $abx + bcy + acz$ runs through a complete residue system modulo abc.

50. Do the previous problem with the word "complete" replaced by "reduced."

51. An integer n is said to be **square free** in case there is no prime p such that $p^2 \mid n$. Show that if $k > 0$ there exist k consecutive integers that are not square free.

4.3 The Theorems of Fermat and Euler

How Powers Fall into Congruence Classes

Assume a and b are relatively prime. Let us examine the powers of $a \pmod b$. Taking $a = 2$ and $b = 7$, for example, we have

n	1	2	3	4	5	6	7
a^n	**2**	**4**	$8 \equiv \mathbf{1}$	$16 \equiv \mathbf{2}$	$32 \equiv \mathbf{4}$	$64 \equiv \mathbf{1}$	$128 \equiv \mathbf{2}$.

Here all the congruences are modulo 7. We have done more work in the above table than we needed to, since, for instance, once we complete the table up to $2^3 \equiv 1$, we have $2^4 \equiv 2^3 \cdot 2 \equiv 1 \cdot 2 \equiv 2$, $2^5 \equiv 2^3 \cdot 2^2 \equiv 4$, etc. Clearly the pattern of least residues $2, 4, 1, 2, 4, 1, \ldots$ will repeat forever.

In the general case, since there are only finitely many congruence classes modulo b, eventually two powers of a will have to fall into the same class. But if

$$a^i \equiv a^j \pmod b, \tag{4.4}$$

then

$$a^{i+1} \equiv a^{j+1} \pmod b$$

follows from multiplication by a; and the pattern is repeated.

Let us look at this process more closely. Suppose (4.4) holds with $i < j$. This can be written

$$a^i \cdot 1 \equiv a^i a^{j-i} \pmod b.$$

Now $(a^i, b) = 1$ by Theorem 1.22, and so by the cancellation theorem

$$a^{j-i} \equiv 1 \pmod b.$$

We see that some power of a must be congruent to 1 modulo b.

DEFINITION. the order of a modulo b

Let a and $b > 0$ be relatively prime integers. By the **order of a modulo b** we mean the least positive integer k such that

$$a^k \equiv 1 \pmod b.$$

Examples. From our previous calculation the order of 2 (mod 7) is 3. The following table shows the successive powers a^n of $a = 1, 2, 3, 4, 5, 6 \pmod 7$ until we hit 1.

a \ n	1	2	3	4	5	6	order
1	1						1
2	2	4	1				3
3	3	2	6	4	5	1	6
4	4	2	1				3
5	5	4	6	2	3	1	6
6	6	1					2

In the table, the powers have been replaced by congruent elements of the reduced residue system $1, 2, 3, 4, 5, 6$; for example, $3^2 = 9 \equiv 2 \pmod 7$. ◇

Suppose a has the order $k \pmod b$. Then if m is any multiple of k, say $m = kt$, we have

$$a^m = (a^k)^t \equiv 1^t \equiv 1 \pmod b.$$

Conversely, if $a^n \equiv 1 \pmod b$, use the division algorithm to write

$$n = kq + r, \quad 0 \le r < k.$$

Then

$$a^n = a^{kq+r} = (a^k)^q a^r \equiv a^r \equiv 1 \pmod b.$$

But this contradicts the assumption that k is the smallest positive integer such that $a^k \equiv 1 \pmod b$ unless $r = 0$. Thus we see that $k \mid n$.

We have already seen that if $a^i \equiv a^j \pmod b$ with $i < j$, then

$$a^{j-i} \equiv 1 \pmod b,$$

and so $k \mid j - i$. In particular, no two powers of a up to the kth power can be congruent modulo b.

We sum up what we have proved in the following theorem.

Theorem 4.6. *Suppose a and $b > 0$ are relatively prime integers, and let k be the order of a modulo b. Then the numbers*

$$a^1, a^2, a^3, \ldots, a^k$$

are incongruent modulo b and every positive power of a is congruent to one of them. Furthermore if m is a positive integer then $a^m \equiv 1 \pmod b$ if and only if $k \mid m$.

Euler's Theorem

We found that the six elements of a reduced residue system modulo 7 had orders as follows:

element	1	2	3	4	5	6
order	1	3	6	3	6	2

Notice that the orders 1, 2, 3, and 6, are all divisors of $6 = \phi(7)$.

If we take $b = 15$, then a reduced residue system has $\phi(15) = 8$ elements. The reader should confirm that the following table gives the correct orders of these elements.

element	1	2	4	7	8	11	13	14
order	1	4	2	4	4	2	4	2

Here the orders are 1, 2, and 4, all divisors of 8. In each case every order is a divisor of $\phi(b)$. It appears that it may be true that the order of any element a modulo b is always a divisor of $\phi(b)$. In light of the last theorem, this is equivalent to saying that

$$a^{\phi(b)} \equiv 1 \pmod{b}$$

whenever $(a, b) = 1$.

In order to prove this it seems natural to try to associate the number a with a reduced residue system modulo b in some way, since the latter has exactly $\phi(b)$ elements. Theorem 4.5 suggests multiplication; if $x_1, x_2, \ldots, x_t$ form a reduced residue system modulo b (where $t = \phi(b)$), then so do $ax_1, ax_2, \ldots, ax_t$.

In particular, the numbers $ax_1, ax_2, \ldots, ax_t$ are congruent in some order to the numbers $x_1, x_2, \ldots, x_t$. We do not know exactly how these elements match up, but certainly if we multiply all the congruences together and sort things out, we get

$$ax_1ax_2 \cdots ax_t \equiv x_1x_2 \cdots x_t \pmod{b},$$

or

$$a^{\phi(b)}(x_1x_2 \cdots x_t) \equiv x_1x_2 \cdots x_t \pmod{b}.$$

By Theorem 1.22 the product of the integers x_i is relatively prime to b and so the cancellation theorem yields

$$a^{\phi(b)} \equiv 1 \pmod{b}.$$

Theorem 4.7 (Euler's theorem). *If a and $b > 0$ are relatively prime integers, then*

$$a^{\phi(b)} \equiv 1 \pmod{b}.$$

Combining Euler's theorem with the last line of Theorem 4.6, we get the following result.

Theorem 4.8. *If a and $b > 0$ are relatively prime integers and if k is the order of a modulo b, then $k \mid \phi(b)$.*

Applications of Euler's Theorem

Let us use Euler's theorem to find the last digit of 17^{102}. What we are after is the least residue of $17^{102} \pmod{10}$. Now $\phi(10) = 4$, so by Euler's theorem

$$17^4 \equiv 1 \pmod{10}.$$

Then

$$17^{102} = 17^{4\cdot 25+2} = (17^4)^{25}17^2 \equiv 1^{25}7^2 = 49 \equiv 9 \pmod{10}.$$

Thus the last digit is 9.

Now suppose we want the last *two* digits of 17^{102}. This will be the least residue of 17^{102} (mod 100). Since $\phi(100) = 40$, Euler's theorem tells us that

$$17^{102} = 17^{40\cdot 2+22} = (17^{40})^2 17^{22} \equiv 1^{40} 17^{22} \equiv 17^{22} \pmod{100}.$$

Although we have reduced the exponent from 102 to 22, this still leaves us with a nasty computation. The integer 17^{22} has 28 digits, and an ordinary calculator will give it in scientific notation, obscuring the last two digits. Of course we could compute $17, 17^2 = 189 \equiv 89, 17^3 = 17 \cdot 17^2 \equiv 17 \cdot 89 = 1513 \equiv 13$, etc., reducing modulo 100 as we go until we hit 17^{22}, but this would be tedious. In the next section, we will give an efficient method of doing such a calculation.

Fermat's Theorem

The case of Euler's theorem when the modulus is a prime p is attributed to Fermat. Actually Fermat merely told people he had proved the theorem bearing his name; Euler first published a proof. Since $\phi(p) = p-1$ we have the following theorem.

Theorem 4.9 (Fermat's theorem). *If p is prime and a is an integer such that $p \nmid a$, then*

$$a^{p-1} \equiv 1 \pmod{p}.$$

Of course, since p is prime, $(a, p) = 1$ is equivalent to $p \nmid a$. A slight variation allows this theorem to be stated with a simpler hypothesis.

Theorem 4.10 (Fermat's theorem, second form). *If p is prime, then*

$$a^p \equiv a \pmod{p}$$

for all integers a.

Proof. If $p \nmid a$, then multiplying the congruence of the first form of the theorem by a gives the conclusion. But if $p \mid a$, then both sides of the new congruence are congruent to 0 modulo p. □

Wilson's Theorem

The argument used to prove Euler's theorem above is too good to let alone. Recall that we derived the congruence

$$x_1 x_2 \cdots x_t \equiv a x_1 a x_2 \ldots a x_t \pmod{b},$$

where $(a, b) = 1$, from the fact that if $x_1, x_2, \ldots, x_t$ was a reduced residue system modulo b, then so was $ax_1, ax_2, \ldots, ax_t$.

Matching things up is the germ of this proof, i.e., matching elements x with congruent elements ax. Let us try to evaluate (mod b) the product $x_1 x_2 \ldots x_t$ itself by a similar argument.

By Theorem 4.2 we know that for each x in a reduced residue system (mod b) there exists a unique element x' in the system such that $xx' \equiv 1 \pmod{b}$. Our idea will be to pair up the elements x in this way, each pair multiplying to $1 \pmod{b}$.

We seem to have proved that $x_1 x_2 \cdots x_t \equiv 1 \pmod{b}$, but a little care is needed. We must consider the possibility that x pairs with itself, that is, $xx \equiv 1 \pmod{b}$. This is certainly the case if $x \equiv \pm 1 \pmod{b}$, and perhaps for other values of x. For example, $4 \cdot 4 \equiv 1 \pmod{15}$.

If we assume the modulus is a prime p we simplify the situation, since if

$$x^2 \equiv 1 \mod p,$$

then $p \mid x^2 - 1 = (x-1)(x+1)$. Then Theorem 2.4 says that p divides one or the other of $x - 1$ and $x + 1$, and so

$$x \equiv \pm 1 \pmod{p}.$$

A prime modulus also has the advantage of a very explicit reduced residue system, namely, $1, 2, \ldots, p-1$.

If $p = 11$, for example, we can pair off everything except 1 and 10 ($\equiv -1$). Indeed,

$$2 \cdot 6 \equiv 3 \cdot 4 \equiv 5 \cdot 9 \equiv 7 \cdot 8 \equiv 1 \pmod{11}.$$

Thus

$$1 \cdot 2 \cdot 3 \cdot 4 \cdot 5 \cdot 6 \cdot 7 \cdot 8 \cdot 9 \cdot 10 = 1(2 \cdot 6)(3 \cdot 4)(5 \cdot 9)(7 \cdot 8)10 \equiv -1 \pmod{11}.$$

The same argument works for any odd prime p, producing a theorem that bears John Wilson's name, even though there is no evidence that Wilson (1741–1793) did more than guess it from numerical evidence. The first published proof was by Lagrange.

Theorem 4.11 (Wilson's theorem). *If p is prime, then*

$$(p-1)! \equiv -1 \pmod{p}.$$

The statement (without proof) of this theorem was first published by the English mathematician Edward Waring in 1770 in his *Meditationes Algebraicae*, along with two other famous conjectures based on numerical evidence. One of these, known as **Waring's problem**, says that each positive integer is the sum of at most 4 squares, of at most 9 cubes, etc. That is, given a positive integer k there exists a least integer $g(k)$ such that each positive integer is the sum of at most $g(k)$ kth powers. A proof that $g(2) = 4$ appears in Section 7.3. (Lagrange first proved this.) Waring's problem was not settled in general until 1909, when the German mathematician David Hilbert proved it. (Hilbert did not give a formula for $g(k)$, but merely showed it always existed.)

The other conjecture, due to Christian Goldbach, says that each even integer greater than 2 can be written as the sum of two primes. This still unproved **Goldbach conjecture** is discussed in Chapter 0.

Problems for Section 4.3.

A

In the first four problems, find the order of each element in the given reduced residue system S (mod b).

1. $b = 5, S = \{1, 2, 3, 4\}$
2. $b = 10, S = \{1, 3, 7, 9\}$
3. $b = 11, S = \{1, 2, \ldots, 10\}$
4. $b = 13, S = \{1, 2, \ldots, 12\}$

In the next six problems, find the order of a (mod b). Compute $\phi(b)$.

5. $a = 3, b = 16$
6. $a = 5, b = 17$
7. $a = 25, b = 18$
8. $a = 18, b = 25$
9. $a = 10, b = 19$
10. $a = 10, b = 21$

In the next eight problems, compute the least residue of a^k (mod b).

11. $a = 2, k = 35, b = 11$
12. $a = 3, k = 45, b = 13$
13. $a = 50, k = 40, b = 19$
14. $a = 60, k = 75, b = 19$
15. $a = 8, k = 50, b = 21$
16. $a = 9, k = 43, b = 25$
17. $a = 51, k = 100, b = 17$
18. $a = 105, k = 77, b = 53$

19. Find the last two digits of 7^{125}.
20. Find the last two digits of 9^{203}.

In the next four problems, find the least residue of $(b - 1)!$ (mod b).

B

21. $b = 9$
22. $b = 10$
23. $b = 101$
24. $b = 21$
25. Prove or disprove: If $b > 1$ is composite, then $(b - 1)! \equiv 0 \pmod{b}$.
26. Show how the matching in the proof of Wilson's theorem goes for $p = 13$.
27. Repeat the previous problem for $p = 17$.
28. Find the least residue of $100! + 102! \pmod{101}$.
29. Find the least residue of $95! \pmod{97}$.
30. Show that if $b > 1$ is not prime, then $(b - 1)! \not\equiv -1 \pmod{b}$.
31. Let $a, b > 0$, and $c > 0$ be integers such that $(a, c) = 1$ and $b \mid c$. Show that the order of a (mod b) divides the order of a (mod c).
32. Show that if $(a, b) = 1$ and i and j are nonnegative integers, then $a^i \equiv a^j \pmod{b}$ if and only if $i \equiv j \pmod{k}$, where k is the order of a (mod b).
33. Give an example of positive integers k and b such that $k \mid \phi(b)$ and yet no integer has order k (mod b).
34. Show that if $(a, b) = 1$ and if $c \equiv a^k \pmod{b}$ for some positive integer k, then the order of c (mod b) does not exceed the order of a (mod b).
35. Show that no integer has order 40 modulo 100.
36. Show that no integer has order 24 modulo 72.
37. Show that if $x^2 \equiv y^2 \pmod{p}$, p a prime, then $x \equiv \pm y \pmod{p}$.
38. Show how the proof of Wilson's theorem breaks down for $p = 2$.
39. Show that if $(a, b) = 1$, then one solution to $ax \equiv c \pmod{b}$ is $x = ca^{\phi(b)-1}$.
40. Show that if $b > 2$, then there exists an integer of order 2 modulo b.
41. Show that $\phi(n)$ is even for $n > 2$.
42. Prove that $g(2) > 3$, where g is as defined at the end of this section.

43. Show that if p is prime, then $(p-2)! \equiv 1 \pmod{p}$.
44. Show that if p is an odd prime, then $(p-3)! \equiv (p-1)/2 \pmod{p}$.
45. Show that if p is an odd prime and $p \nmid a$, then $a^{(p-1)/2} \equiv \pm 1 \pmod{p}$.
46. We call p a **twin prime** if p is prime and also $p-2$ or $p+2$ is prime. Find the smallest even integer greater than 4 that is not the sum of two twin primes. (The two primes do not have to be twins of each other.)

C

In the next two problems, assume a has order k (mod p), p a prime. Assume $d > 0$.

47. Show that if $(d, k) = 1$, then a^d has order $k \pmod{p}$.
48. Show that if $(d, k) > 1$, then a^d has order $< k \pmod{p}$.
49. Suppose that a has order $p-1 \pmod{p}$, p a prime, and $j > 0$. Show that the order of $a^j \pmod{p}$ is $(p-1)/(j, p-1)$.
50. Assume the hypotheses of the previous problem. Show that if $k \mid p-1$, $k > 0$, then any reduced residue system modulo p contains exactly $\phi(k)$ elements of order $k \pmod{p}$.
51. Prove Fermat's theorem (second version) for positive integers a by induction on a, using the binomial theorem. (A statement of the binomial theorem may be found in the exercises for Section 5.2.)

4.4 Primality Testing

The Contrapositive of Fermat's Theorem

By Fermat's theorem, if n is prime and $n \nmid a$, then $a^{n-1} \equiv 1 \pmod{n}$. Thus if $a^{n-1} \not\equiv 1 \pmod{n}$, n cannot be prime. This idea has interesting consequences.

Pretend we do not know if 33 is prime or not. If it were prime, then since $33 \nmid 2$ we would have $2^{32} \equiv 1 \pmod{33}$. But in fact,

$$2^{32} = (2^5)^6 2^2 = 32^6 2^2 \equiv (-1)^6 2^2 = 4 \pmod{33}.$$

Thus we have proved 33 is not prime, *without exhibiting a factor between* 1 *and* 33.

This looks like a promising way to distinguish primes from composites, although, of course, it is conceivable that n could be composite and still satisfy the conclusion of Fermat's theorem. Since telling which even integers are prime is easy enough, let us look at odd values of n. We take $a = 2$ for simplicity, and compute the least residue r of $2^{n-1} \pmod{n}$.

n	3	5	7	9	11	13	15	17	19	21	23	25	27	29	31	33
r	1	1	1	4	1	1	4	1	1	4	1	16	13	1	1	4

From this limited evidence we might conjecture that the converse of Fermat's theorem is also true, at least for $a = 2$, that is, that if $2^{n-1} \equiv 1 \pmod{n}$, then n is prime. Even if this is correct, however, it would be of limited use in determining the primality of large values of n without a more efficient way to compute 2^{n-1}. The computation of $2^{32} \pmod{33}$ above was facilitated by noticing that 33 was exactly 1 more than 32, a power of 2. This trick will not work in general.

Our point of view is that n is so large that we do not know if it is prime or not. Thus we cannot use Euler's theorem to simplify the calculation of 2^{n-1}, since computing $\phi(n)$ in any efficient way requires knowing the factorization of n into primes, and that is precisely what we do not know!

We can alway go back to deciding if the odd integer n is prime by checking it for divisiblity by odd integers $\leq \sqrt{n}$, by Theorem 2.8. (Although we really only need to check possible *prime* divisors, determining whether a large possible divisor is prime or not is more work than just dividing it into n.) This entails about $\sqrt{n}/2$ divisions. Thus any scheme to tell if n is prime needs fewer steps than this to be worth considering. Computing $2^{n-1} \pmod{m}$ by starting with 2 and multiplying $n-2$ times by 2, reducing modulo n as we go, takes $n-2$ multiplications and the same number of divisions, and so is not acceptable.

Modular Exponentiation

Fortunately there is a much more efficient way of computing powers modulo n. Suppose we wish to compute $a^d \pmod{n}$. As an example, we will compute the least residue of $848^{187} \pmod{1189}$, so that $a = 848$, $d = 187$, and $n = 1189$.

This seems to be a formidible task, but we will show how to do it using only a hand calculator.

We start by converting the exponent d to its base 2, or binary, representation. An easy way to do this is to successively divide d by 2, keeping track of the remainders (all of which are 0 or 1). For $d = 187$ we have

$$\begin{aligned} 187 &\equiv 93 \cdot 2 + 1 \\ 93 &\equiv 46 \cdot 2 + 1 \\ 46 &\equiv 23 \cdot 2 + 0 \\ 23 &\equiv 11 \cdot 2 + 1 \\ 11 &\equiv 5 \cdot 2 + 1 \\ 5 &\equiv 2 \cdot 2 + 1 \\ 2 &\equiv 1 \cdot 2 + 0 \\ 1 &\equiv 0 \cdot 2 + 1. \end{aligned}$$

Then the remainders, listed in reverse order, give the binary representation of d. In our example

$$\begin{aligned} d &= 187 = 10111011_2 \\ &= 1 \cdot 2^7 + 0 \cdot 2^6 + 1 \cdot 2^5 + 1 \cdot 2^4 + 1 \cdot 2^3 + 0 \cdot 2^2 + 1 \cdot 2^1 + 1 \\ &= 128 + 32 + 16 + 8 + 2 + 1. \end{aligned}$$

Now in order to compute a to the power $d \pmod{n}$ we use the calculator to successively square and reduce modulo n as follows. (A method for finding least residues with a hand calculator is given at the end of Section 1.2.)

k	$a^k \pmod{n}$
1	848
2	$848^2 \equiv 719{,}104 \equiv 948 \pmod{1189}$
4	$948^2 \equiv 898{,}704 \equiv 1009 \pmod{1189}$
8	$1009^2 \equiv 1{,}018{,}081 \equiv 297 \pmod{1189}$
16	$297^2 \equiv 88{,}209 \equiv 223 \pmod{1189}$
32	$223^2 \equiv 49{,}729 \equiv 980 \pmod{1189}$
64	$980^2 \equiv 960{,}400 \equiv 877 \pmod{1189}$
128	$877^2 \equiv 769{,}129 \equiv 1035 \pmod{1189}$

Then we have

$$\begin{aligned} 848^{187} &= 848^{128+32+16+8+2+1} \\ &= 848^{128} 848^{32} 848^{16} 848^{8} 848^{2} 848^{1} \\ &\equiv 1035 \cdot 980 \cdot 223 \cdot 297 \cdot 948 \cdot 848 \pmod{1189}. \end{aligned}$$

By multiplying out the last expression factor by factor, reducing modulo 1189 as we go, we find

$$848^{187} \equiv 190 \pmod{1189}.$$

Example. Compute the least residue of $2^{340} \pmod{341}$ using this method.

The reader should check the details of this example with a calculator or computer. We find $340 = 101010100_2 = 256 + 64 + 16 + 4$. Now $2^1 \equiv 2$, $2^2 \equiv 4$, $2^4 \equiv 16$, $2^8 \equiv 256$, $2^{16} \equiv 64$, $2^{32} \equiv 4$, $2^{64} \equiv 16$, $2^{128} \equiv 256$, and $2^{256} \equiv 64$, all modulo 341. Thus

$$2^{340} = 2^{256+64+16+4} = 2^{256}2^{64}2^{16}2^4 \equiv 64 \cdot 16 \cdot 64 \cdot 16 \equiv 1 \pmod{341}.$$

Complexity Theory

To explain what we mean in saying the above modular exponentiation algorithm is "efficient," we must talk about *computational complexity*, a subject that has become important because of computers. We would like to estimate the number of steps a given algorithm takes. Actually what we will count will be the elementary operations of addition, subtraction, multiplication, division, and comparison. Although the size of the numbers involved may affect how long such an operation takes on a computer, to keep things simple we will assume each such operation takes the same amount of time, say one-billionth of a second.

Of course, the number of elementary operations in an algorithm depends on the size of the problem to which it is applied. In the case of the modular exponentiation algorithm, this size is measured by d, the exponent. We start by converting d to binary by divisions by 2. Each division determines a binary digit. Suppose there are k of these, namely, $a_0, a_1, \ldots, a_{k-1}$, so that $d = a_{k-1}2^{k-1} + a_{k-2}2^{k-2} + \cdots + a_1 2 + a_0$ with $a_{k-1} = 1$. Note that $d \geq 2^{k-1}$. This part of the algorithm requires k divisions. There are also k comparisons, since after each division we must check if the quotient is 0 to decide when to stop.

Now we compute $d^{2^1}, d^{2^2}, d^{2^3}, \ldots, d^{2^{k-1}} \pmod{n}$ by successively squaring and reducing modulo n (i.e., dividing by n and taking the remainder). This accounts for $k-1$ multiplications and $k-1$ divisions.

Finally, we have at most k integers whose product we must compute (mod n). Needed are at most $k-1$ more multiplications and $k-1$ divisions.

Our analysis has accounted for $2k + 2(k-1) + 2(k-1) = 6k - 4$ elementary operations. We had $2^{k-1} \leq d$, so $k - 1 = \log_2 2^{k-1} \leq \log_2 d$, which implies $6k - 4 \leq 2 + 6\log_2 d$.

Theorem 4.12. *Let a, $d > 0$, and $n > 0$ be integers. Then the least residue of a^d modulo n can be computed with no more than $2 + 6\log_2 d$ elementary operations.*

In the next section, we will consider computing $a^d \pmod{n}$ with $d \approx 10^{300}$. This could be done with no more than $2 + 6\log_2 10^{300} = 2 + 6 \cdot 300\log_2 10 \approx 5981$ elementary operations. (Note that $\log_2 10 = 1/\log_{10} 2 \approx 3.322$.) Compare this with computing $a^2, a^2, \ldots, a^d$, reducing modulo n after each multiplication. There would be about $2 \cdot 10^{300}$ elementary operations, and the sun would burn out before a computer doing one billion per second could finish.

Lamé's Theorem

Interest in how many steps an algorithm takes predates computers. The following theorem was proved in 1844 by Gabriel Lamé.

Theorem 4.13. *If the Euclidean algorithm is applied to two positive integers, then the number of divisions will not exceed 5 times the number of decimal digits of the smaller.*

Proof. Assume $0 < a < b$ and let the Euclidean algorithm be applied as in Section 1.3.

$$\begin{array}{rlc} b = aq_1 + r_1, & & 0 < r_1 < a, \\ a = r_1q_2 + r_2, & & 0 < r_2 < r_1, \\ r_1 = r_2q_3 + r_3, & & 0 < r_3 < r_2, \\ \vdots & & \vdots \\ r_{n-2} = r_{n-1}q_n + r_n, & & 0 < r_n < r_{n-1}, \\ r_{n-1} = r_nq_{n+1}. & & \end{array}$$

Note that $q_i \geq 1$ for $1 \leq i \leq n$, while $q_{n+1} \geq 2$, since $q_{i+1} = 1$ implies that $r_{n-1} = r_n$.

This proof will use the Fibonacci numbers $f_1 = 1, f_2 = 1, f_3 = 2, \ldots$. Since r_n is the last nonzero remainder,

$$r_n \geq 1 = f_2.$$

Because $q_{n+1} \geq 2$ we have

$$r_{n-1} \geq 2r_n \geq 2 = f_3.$$

Likewise, using the Euclidean algorithm equations and $q_i \geq 1$ we have

$$\begin{array}{rcl} r_{n-2} & \geq & r_{n-1} + r_n \geq f_3 + f_2 = f_4, \\ r_{n-3} & \geq & r_{n-2} + r_{n-1} \geq f_4 + f_3 = f_5, \end{array}$$

and, in general, $r_{n-t} \geq f_{t+2}$. Taking t to be $n-2$ and $n-1$ gives

$$r_2 \geq f_n \quad \text{and} \quad r_1 \geq f_{n+1}.$$

Thus

$$a \geq r_1 + r_2 \geq f_{n+1} + f_n = f_{n+2}.$$

From this and Theorem 2.2 we have $a > A^n$, where $A = (\sqrt{5}+1)/2$. Note that $\log_{10} A > \frac{1}{5}$. Suppose a has k decimal digits, so that $a < 10^k$. Then

$$k = \log_{10} 10^k > \log_{10} a > \log_{10} A^n = n \log_{10} A > \frac{n}{5}.$$

Thus $n < 5k$. Since n and $5k$ are integers, $n+1 \leq 5k$. This concludes the proof, since $n+1$ is the number of divisions in the Euclidean algorithm. □

Pseudoprimes

Now we return to the question of whether

$$2^{n-1} \equiv 1 \pmod{n} \tag{4.5}$$

is not only a necessary condition for n to be prime, by Fermat's theorem, but is also sufficient. The ancient Chinese believed this to be true.

The question has already been answered—did you catch it? In an example earlier in this section illustrating the modular exponentiation algorithm, we computed

$$2^{340} \equiv 1 \pmod{341}.$$

This has the form of (4.5) with $n = 341$. But 341 is not prime: $341 = 11 \cdot 31$. Thus (4.5) can be used to identify composite numbers but not primes. Nonetheless, numbers like 341 are rare.

DEFINITION. pseudoprime, pseudoprime to base a

We call a n a **pseudoprime** if $2^{n-1} \equiv 1 \pmod{n}$, but n is composite. More generally, a composite number n such that $a^{n-1} \equiv 1 \pmod{n}$ is called a **pseudoprime to base** a.

The smallest pseudoprime is 341, and was not discovered until 1819, so the Chinese could be excused for their assumption. Of course, bases other than 2 may also be used to identify composite numbers. For example,

$$3^{340} \equiv 56 \pmod{341},$$

providing a factorless proof that 341 is not prime.

Although there are infinitely many pseudoprimes to base 2 (see the problems at the end of this section), they are much rarer than primes. Thus if a randomly chosen integer n satisfies (4.5) it is probably prime. Even rarer are pseudoprimes to multiple bases. For example, there are only 1770 integers below $25 \cdot 10^9$ that are simultaneously pseudoprimes to the bases 2, 3, 5, and 7. Thus the primality of numbers less than $25 \cdot 10^9$ could be determined by testing Fermat's congruence with these four bases, then comparing any number passing all four tests with a list of the 1770 exceptions.

We might hope that for any composite number n, there is some base a for which Fermat's theorem could be used to show that n is composite. We hope in vain; there are composite integers, called **Carmichael numbers**, which are pseudoprimes to every base. That is, n is composite, but

$$a^{n-1} \equiv 1 \pmod{n}$$

whenever $(a, n) = 1$. The smallest is $561 = 3 \cdot 11 \cdot 17$. It was only proved in 1994, by Alford, Granville, and Pomerance, that there are infinitely many Carmichael numbers. Their proof was based on a suggestion of Paul Erdös.

Mersenne and Fermat Numbers

By using Fermat's theorem with multiple bases to weed out most composite numbers, and then more sophisticated tests, the primality of numbers of up to 150 digits can be determined in a few seconds with a computer. For integers of a special form, such as Mersenne and Fermat numbers, even better methods are available, enabling the primality of far larger numbers to be determined.

The **Lucas-Lehmer test**, which has been used to identify many Mersenne primes, is a century-spanning theorem, since the 1878 test of the Frenchman Edouard Lucas was simplified by the American D. H. Lehmer in 1930. We define a sequence $S_1, S_2, \ldots$ by $S_1 = 4$, and $S_n = S_{n-1}^2 - 2$ for $n > 1$. For example, $S_2 = 4^2 - 2 = 14$ and $S_3 = 14^2 - 2 = 194$. The test says that if p is an odd prime, then $M_p = 2^p - 1$ is prime if and only if $S_{p-1} \equiv 0 \pmod{M_p}$.

As an example, take $p = 7$, so $M_p = 2^7 - 1 = 127$. Then $S_1 = 4, S_2 = 14, S_3 = 194 \equiv 67, S_4 \equiv 67^2 - 2 = 4487 \equiv 42, S_5 \equiv 42^2 - 2 = 1762 \equiv 111$, and $S_6 \equiv 111^2 - 2 = 12319 \equiv 0$, with all congruence modulo 127. This proves that 127 is prime.

An analogous test for Fermat numbers is the following.

Theorem 4.14 (Pepin's test). *If $n > 0$, the Fermat number $F_n = 2^{2^n} + 1$ is prime if and only if*

$$3^{(F_n-1)/2} \equiv -1 \pmod{F_n}. \tag{4.6}$$

Proof. We will only prove the "if" part here, but a proof of the "only if" part is in the problems for Section 5.4. Assume (4.6). Then if p is any prime dividing F_n we have

$$3^{(F_n-1)/2} \equiv -1 \pmod{p}$$

by part (6) of Theorem 1.16 (so $p \neq 3$), and squaring gives

$$3^{F_n-1} \equiv 1 \pmod{p}.$$

Let k be the order of 3 modulo p. Theorem 4.6 says that k divides $F_n - 1 = 2^{2^n}$. Thus $k = 2^t$ for some integer $t \leq 2^n$. Suppose $t < 2^n$. Then we can raise both sides of the congruence $3^k \equiv 1 \pmod{p}$ to the power $2^{2^n-t-1} \geq 1$ to get

$$1 \equiv (3^k)^{2^{2^n-t-1}} = 3^{2^t(2^{2^n-t-1})} = 3^{2^{2^n-1}} = 3^{2^{2^n}/2} = 3^{(F_n-1)/2} \equiv -1 \pmod{p}.$$

But this means $p = 2$, which is impossible.

We must have $t = 2^n$ and $k = 2^{2^n} = F_n - 1$. Now by Fermat's theorem $k \leq p - 1$. Thus $p \geq k + 1 = F_n$. Since p is a divisor of F_n, we must have $p = F_n$. Thus F_n is prime. □

Example. Use Pepin's test to show that $F_3 = 257$ is prime.

By the part of Pepin's test proved above, it suffices to show that $3^{(257-1)/2} = 3^{128} \equiv -1 \pmod{257}$. But $3^2 \equiv 9$, $3^4 \equiv 81$, $3^8 \equiv 81^2 \equiv 136$, $3^{16} \equiv 136^2 \equiv 249$, $3^{32} \equiv 249^2 \equiv 64$, $3^{64} \equiv 64^2 \equiv 241$, and $3^{128} \equiv 241^2 \equiv 256 \equiv -1$, with all congruences modulo 257. ◇

As we have seen, Fermat's theorem may tell us that a number is composite without providing a factor. In fact, factoring appears to be much harder (in terms of computation time) than determining primality. Although methods have been developed that are considerably better than trying all divisors up to the square root of the number to be factored, they are not powerful enough to factor, say, an arbitrary 300-digit number in any reasonable amount of time.

In 1994 a group at Bellcore in Redbank, New Jersey, announced the factorization of a 129-digit number into two huge primes. The feat, which took 8 months and was aided by the computers of 600 Internet volunteers, may have been the largest computation ever at the time. Factoring the number had been set as a seemingly impossible task in a 1977 *Scientific American* column by Martin Gardner.

Problems for Section 4.4.

In the first four problems, convert to binary.

1. 53
2. 72
3. 111
4. 145

In the next four problems, find the least residue of $117^d \pmod{1081}$.

5. $d = 53$
6. $d = 72$
7. $d = 111$
8. $d = 145$

In the next six problems, use the modular exponentiation method of this section.

9. Find the least residue of $2^{32} \pmod{33}$.
10. Confirm that $3^{340} \equiv 56 \pmod{341}$.
11. Find the least residue of $35^{100} \pmod{53}$.
12. Find the least residue of $619^{55} \pmod{733}$.
13. Use Fermat's theorem with $a = 2$ to show that 91 is composite.
14. Use Fermat's theorem with $a = 5$ to show that 143 is composite.

B

In the next two problems, let $S_1, S_2, \ldots$ *be as in the Lucas-Lehmer test.*

15. Compute the least residues of $S_1, S_2, \ldots, S_{10} \pmod{M_{11}}$. What do you conclude?
16. Compute the least residues of $S_1, S_2, \ldots, S_{12} \pmod{M_{13}}$. What do you conclude?

17. According to Pepin's test, what congruence must be satisfied to show that F_2 is prime?
18. What is the least residue of $3^{2^{15}}$ (mod 65,537)?
19. Suppose $(a, 561) = 1$. Show that $a^{560} \equiv 1 \pmod{k}$ for $k = 3$, 11, and 17.
20. Prove that if $(a, 561) = 1$, then $a^{560} \equiv 1 \pmod{561}$, thus proving that 561 is a Carmichael number.
21. Show that if n is an odd pseudoprime to base a, then n is a pseudoprime to base $-a$.
22. Prove that there are infinitely many pseudoprimes to base 1, and also to base -1.

23. Prove that if n is a pseudoprime to base a with $|a| > 1$, then n is a pseudoprime to infinitely many bases.
24. Prove that if n is a pseudoprime to base a and to base b, then n is a pseudoprime to base ab.

C 25. Prove that 1729 is a Carmichael number.

26. Suppose that $n = p_1 p_2 \cdots p_t$, where $t > 1$ and $p_1, p_2, \ldots, p_t$ are distinct primes such that $p_i - 1 \mid n - 1$ for $i = 1, 2, \ldots, t$. Show that n is a Carmichael number.

The next four problems prove that there exist infinitely many pseudoprimes.

27. Show that if n is a pseudoprime, then $n \mid M_n - 1$.
28. Show that if $n \mid M_n - 1$, then $M_n \mid 2^{M_n - 1} - 1$. (**Hint**: Theorem 3.7.)
29. Show that if n is an odd pseudoprime, then so is M_n.
30. Show that there exist infinitely many odd pseudoprimes.

4.5 Public-Key Cryptography

The Idea of a Key

Cryptography is a method of sending a message in a form that only its intended recipient can understand. Although this calls to mind spies, diplomats, and the military, cryptography has wider application. In an age when more and more information is transmitted over telephone lines or by radio (and so is subject to interception), keeping one's messages (and credit card number) secret is an increasing problem for everybody.

Like everything else, cryptography has its own special language. The original message is called the **plaintext**, and the (supposedly) unreadable version of it is called the **ciphertext**. The processes of going from plaintext to ciphertext and back are called **enciphering** and **deciphering**, respectively.

Often enciphering methods involve a **key** that is known to the sender and intended receiver of the message but no one else. The idea is that someone who does not know the key will not be able to decipher the message even if he or she knows the general method of encipherment.

We will illustrate the idea of a key with a substitution cipher, one of the oldest and simplest methods. Substitution amounts to merely replacing each letter of the alphabet with another letter. Although any message of a few sentences or more enciphered by substitution may be easily figured out (in fact, such problems appear in newspapers and puzzle magazines), our aim with this example is simply to show the concepts involved.

We will use as our key the words "number theory." First, we cross out any repeated letters, leaving

NUMBERTHOY.

Now we write these letters under the alphabet, followed by the unused letters of the alphabet, in order.

Plaintext:	A	B	C	D	E	F	G	H	I	J	K	L	M	N	O	P	Q	R	S	T	U	V	W	X	Y	Z
Ciphertext:	N	U	M	B	E	R	T	H	O	Y	A	C	D	F	G	I	J	K	L	P	Q	S	V	W	X	Z

To encipher we merely replace each letter of our message with the corresponding ciphertext letter. Suppose our message is SEND MONEY. We would replace the S by L, since L is below S in our table, etc. We get

Plaintext: SEND MONEY
Ciphertext: LEFB DGFEX

Of course to decipher LEFB DGFEX the intended receiver of the message would use the key to make his own table like the one above. He would then use it in reverse, changing L to S, E to E, F to N, etc., to retrieve the original message.

The RSA Method

In any method of encipherment using a secret key, the key itself is a weak link. It must be known to both parties who want to communicate, so a breach in security by either will destroy the secrecy of messages. If the key is so complicated that it must be written down, opportunities for discovering it increase. Also it must be prearranged. Parties wishing to communicate secretly for the first time must first agree on the key by some more secure method, such as a courier.

These reasons have led to the development of **public-key cryptography**. Such a system uses a key that allows encipherment but not decipherment of messages. Each party then publishes his own key for anyone who wants to communicate with him to use. He keeps secret additional information that allows him alone to decipher incoming messages.

Systems as we have just described may seem impossible at first, since in traditional enciphering methods knowing how to encipher and decipher go hand in hand. To be accurate, we must admit that in the public-key systems, decipherment by unauthorized parties cannot be said to be impossible. The goal is merely to place such decipherment beyond available resources of time and money. If it would take the world's fastest computer 1000 years using the most efficient methods known to decipher a message, then the message may be considered secure, at least unless better methods are found.

In actual use, the public-key methods we will describe are implemented by computers, both for encipherment and decipherment. Large numbers are used to make unauthorized decipherment, even using computers, impractical. In this section, however, we will scale the systems down to human dimensions, so that encipherment and decipherment are reasonable tasks for a person using a hand calculator.

The method we will describe is called the **RSA method**. The name is taken from the initials of its inventors, R. L. Rivest, A. Shamir, and L. Adleman, of the Massachusetts Institute of Technology. The security of messages enciphered by the method depends on the difficulty of factoring a large number n. This number will be the product of two large primes, p and q. To be safe from unauthorized decipherment using a large computer, p and q should each be about 150 (decimal) digits long, making $n = pq$ a number of about 300 digits. To illustrate the method, however, we will use much smaller numbers, say $p = 29$ and $q = 41$. Then their product is $n = 1189$. We also need a positive integer e that is relatively prime to $\phi(n) = (p-1)(q-1)$. In our example, $\phi(n) = 28 \cdot 40$, and we will take $e = 3$.

Now we "publish" the numbers n and e. This might mean allowing the computer of anyone who wants to communicate with us to access n and e from our website. In our example, our public key is the pair $n = 1189$, $e = 3$. We keep the numbers p and q secret, not even revealing them to those from whom we expect secret messages.

RSA Enciphering

Suppose someone wants to send us a message no one else can read. She first converts the message to numerical form, and breaks the resulting sequence of digits into blocks having fewer digits than n. In our example, we will use the following simple conversion.

symbol	number	symbol	number	symbol	number	symbol	number
space	00	G	07	N	14	U	21
A	01	H	08	O	15	V	22
B	02	I	09	P	16	W	23
C	03	J	10	Q	17	X	24
D	04	K	11	R	18	Y	25
E	05	L	12	S	19	Z	26
F	06	M	13	T	20		

Suppose the message to be sent is `SEND MONEY` Then the corresponding sequence of digits is

$$19 \quad 05 \quad 14 \quad 04 \quad 00 \quad 13 \quad 15 \quad 14 \quad 05 \quad 25.$$

Since $n = 1189$ has four digits, our correspondent will break the message into groups of three digits:

$$190 \quad 514 \quad 040 \quad 013 \quad 151 \quad 405 \quad 250.$$

(An extra 0 was added to make the message come out even.)

We label these integers $P_1, P_2, \ldots$. The "P" stands for plaintext; since this method of transforming the message into a sequence of numbers is agreed upon beforehand and public knowledge, anyone could decipher the sequence at this stage.

The enciphered message (or ciphertext) is the sequence $C_1, C_2, \ldots$, where C_i is defined to be the least residue of $P_i^e \pmod{n}$. In our example, C_i is the least residue of $P_i^3 \pmod{1189}$. Since the numbers n and e have been published, anyone who wants to correspond with us can encipher a message this way. (In practice, since the numbers involved are very large, the encipherment is done with a computer.)

In our example, P_1 is 190, and so

$$C_1 \equiv 190^3 = 6{,}859{,}000 \equiv 848 \pmod{1189}.$$

Thus we see $C_1 = 848$. Likewise our correspondent computes C_2 by

$$C_2 \equiv 514^3 = 135{,}796{,}744 \equiv 1054 \pmod{1189}.$$

If necessary she can use the modular exponention method explained in Section 4.4. Continuing in this way she finds $C_1, C_2, \ldots, C_7$ to be

$$848, \quad 1054, \quad 983, \quad 1008, \quad 796, \quad 695, \quad 351,$$

and this is the ciphertext she sends to us.

RSA Deciphering

Having received the above message $C_1, C_2, \ldots$, now it is up to us to decipher it. We must undo the scrambling of the original message caused by raising the numbers $P_1, P_2, \ldots$ to the power e. This we will do by raising each number C_i to the power d, where d will be defined shortly. First we define the integer b by

$$b = [p-1, q-1].$$

Note that someone not knowing p and q will not be able to compute b. Now we define d to be the least positive solution to

$$ex \equiv 1 \pmod{b}.$$

In our example, $b = [28, 40] = 28 \cdot 40/(28, 40) = 28 \cdot 40/4 = 280$, where the second equality follows from Theorem 1.5. Thus the congruence we must solve to find d is

$$3x \equiv 1 \pmod{280}.$$

Using the method of Section 4.1, we easily find the least positive solution to be $x = 187$, so this is our value of d.

Now all we have to do to retrieve the plaintext message is find the least residue of each ciphertext number C_i when raised to the power d. For example P_1 is just the least residue modulo 1189 of

$$C_1^d = 848^{187}.$$

We will certainly need to use the modular exponentiation algorithm to do this calculation. In fact, we already have! Computing $848^{187} \pmod{1189}$ was our first example using the algorithm in Section 4.4, where we found that

$$848^{187} \equiv 190 \pmod{1189}.$$

Note that 190 is the original plaintext number P_1. In the same way, we can calculate the other plaintext numbers 514, 40, 13, 151, 405, and 250 as the least residues of 1054^{187}, 983^{187}, ... modulo 1189. We list $P_1, P_2, \ldots$ as three-place numbers,

$$190 \quad 514 \quad 040 \quad 013 \quad 151 \quad 405 \quad 250,$$

regroup into pairs,

$$19 \quad 05 \quad 14 \quad 04 \quad 00 \quad 13 \quad 15 \quad 14 \quad 05 \quad 25,$$

and, using the space = 00, `A` = 01, correspondence we agreed on, read the message

`S E N D _ M O N E Y.`

The Complexity of the RSA Method

The reader, after plowing through the above calculations, may readily agree that RSA enciphering and deciphering are best left to computers. He or she may also have noticed that the difficulty of factoring our $n = 1189$, upon which the security of the method depends, is much less than that of the other calculations the method entails. Certainly if we can raise 848 to the 187 power modulo 1189, then someone else can factor 1189, and so read secret messages sent to us.

The explanation for this paradox has to do with the size of the numbers involved, and with the complexity of the calculation we must do with these numbers. Recall that in actual practice the primes p and q will be about 150 digits each, and n will be about 300 digits. It turns out that for numbers of this size, the work involved in enciphering and deciphering, while not to be sneezed at, is far less than that of factoring n by any known method.

Let us estimate the complexity of deciphering one ciphertext number. (We choose deciphering instead of enciphering because it is harder, since b and d must also be computed.) We can compute $b = [p-1, q-1]$ as $(p-1)(q-1)/(p-1, q-1)$ by Theorem 1.5. Applying the Euclidean algorithm to calculate $(p-1, q-1)$ takes at most $5 \cdot 150 = 750$ divisions by Lamé's theorem of the previous section. Since we need to check for a 0 remainder after each division to tell when to stop, we will add 750 comparisons, for a total of 1500 elementary operations.

We also need to compute d, which is the smallest positive solution to $ex \equiv 1 \pmod{b}$. In Section 4.1, we learned to solve such linear congruences by applying the Euclidean algorithm to e and b. Whatever e is, $b < pq \approx 10^{300}$. Thus by Lame's theorem again at most $5 \cdot 300 = 1500$ divisions and 1500 comparisons are needed.

Actually the Euclidean algorithm method also entails solving the equations of the algorithm backward to find x and y such that $ex + by = 1$, so that $ex \equiv 1 \pmod{b}$. There is a computationally superior method of finding x, however. Define a sequence $x_{-1}, x_0, x_1, x_2, \ldots, x_N$ by

$$x_{-1} = 0, \quad x_0 = 1, \quad \text{and} \quad x_i = x_{i-2} - q_i x_{i-1} \text{ for } i = 1, 2, \ldots, N, \tag{4.7}$$

where q_i is the ith quotient from the Euclidean algorithm and r_N is the last nonzero remainder (so $N \leq 1500$). Then, as will be shown in the problems at the end of this section,

$$ex_i \equiv r_i \pmod{b} \text{ for } i = 1, 2, \ldots, N.$$

In particular, $ex_N \equiv r_N = 1 \pmod{b}$. Thus a solution $d = x_N$ can be calculated by applying the recursion of (4.7) N times, each requiring a subtraction and multiplication, for no more than $2N \leq 3000$ additional elementary operations.

So far we have accounted for at most 1500 operations to compute b, 3000 to apply the Euclidean algorithm to e and b, and 3000 more to compute d, for a total of 7500 elementary operations. It only remains to find the least residue of $C^d \pmod{n}$. Since d is the least positive solution to a congruence with modulus $b < 10^{300}$, by the discussion after Theorem 4.12 fewer than 6000 elementary operations are needed. Thus the total number of elementary operations needed to decipher one ciphertext number is fewer than 13,500. Any reasonable computer could do this in a fraction of a second. Since computing b and d only needs to be done once, deciphering one million ciphertext numbers would take fewer than $7500+1{,}000{,}000\cdot 6000 \approx 6\cdot 10^9$ operations, which would take about 6 seconds on our hypthetical computer doing one billion per second. This compares with the problem of factoring the 300-digit number n, which appears to be well beyond the range of present computing methods. Trying odd divisors of n up to $\sqrt{n}$ entails about $10^{150}/2$ divisions. At one billion per second, this would take about $1.6\cdot 10^{133}$ years. Better methods are known, but not sufficiently better to make factorization of such large numbers practical.

Actually RSA enciphering and deciphering are slower than comparably secure private-key methods, and so in practice (for example, for secure internet communication) the RSA method is used to exchange a private key (which is changed after each communication), and then further transmission uses the faster conventional system.

Why the Method Works

Why does raising the ciphertext numbers to the power d bring back the plaintext? Consider the plaintext number P. The corresponding ciphertext is $C \equiv P^e \pmod{n}$. We decipher C by computing

$$P' \equiv C^d \pmod{n},$$

and so it suffices to show that $P' = P$. In fact since both P and P' are nonnegative integers less than n, it suffices to show that

$$P' \equiv P \pmod{n}.$$

Now by the definitions of C and P' we have

$$P' \equiv C^d \equiv (P^e)^d \equiv P^{de} \pmod{n}.$$

Recall that d is the least positive solution to

$$ex \equiv 1 \pmod{b},$$

where $b = [p-1, q-1]$. (Since e is chosen relatively prime to $p-1$ and $q-1$, it is relatively prime to b and so the congruence has a solution by Theorem 4.2.)

Let $de = 1 + kb$. Now if $p \nmid P$, then

$$P^{p-1} \equiv 1 \pmod{p}$$

by Fermat's theorem. Then

$$P^{kb} \equiv 1 \pmod{p}$$

since $p-1$ divides b. Thus

$$P^{de} = P^{kb+1} \equiv P \pmod{p},$$

and this congruence is clearly true also if p divides P. In the same way,

$$P^{de} \equiv P \pmod{q}.$$

Thus we have

$$P^{de} \equiv P \pmod{n},$$

since both p and q divide $P^{de} - P$. But recall that $P' \equiv P^{de} \pmod{n}$.

How Secure are Public-Key Systems?

The security of the RSA method depends on the fact that no efficient way is known of factoring large numbers. If a dramatically faster factoring method were found, then the method would become invalid. When another public-key system, called the "knapsack method," was cracked in 1982 by the Israeli mathematician Adi Shamir (the "S" of RSA), *Time* magazine ran a full-page article on the feat. Two years later, a more complicated knapsack scheme was broken by Ernest F. Brickell of the Scandia National Laboratories. (See the next two pages. The second story is from the Boston *Globe*.) For an explanation of the knapsack method see the *Scientific American* article by Hellman listed in the references.

Problems for Section 4.5.

A

Problems 1 through 10 assume the substitution method describe at the beginning of this section. In problems 1 through 8 assume the key is `NUMBER THEORY`.

1. Encipher: `SELL ON MONDAY`
2. Encipher: `SNODGRASS IS A DOUBLE AGENT`
3. Encipher: `CHERCHEZ LA FEMME`
4. Encipher your name.
5. Decipher: `NULPNOF RKGD UENFL`
6. Decipher: `PHOL OL N ROFE UGGA`
7. Decipher: `PHE MCGMA VOCC EWICGBE`
8. Decipher: `REKDNPL CNLP PHEGKED`
9. Let the key be `LAGRANGE`.

   Encipher: `EVERY INTEGER IS THE SUM OF FOUR SQUARES.`

   Decipher: `TCN TCNMQNJ ME WDISMK DS TQUN.`
10. Let the key be `EUCLID`.

    Encipher: `THERE ARE INFINITELY MANY PRIMES.`

    Decipher: `SBIQI FR MN QNYEJ QNEL SN AINKISQY.`

Problems 11 through 34 refer to the RSA method. In the next four problems, change each message to a sequence $P_1, \ldots, P_4$ *using the system of this section.*

TIME

THE WEEKLY NEWSMAGAZINE

Computers

Opening the "Trapdoor Knapsack"

An Israeli mathematician cracks a formidable code

Five years ago, computer scientists at Stanford and M.I.T. made a pair of chummy but keenly competitive $100 bets. A team at each university had devised a secret code to protect computers from electronic intruders by scrambling and unscrambling the data in a complex fashion. Each team offered cash to the first mathematician who could crack its code, figuring that the deciphering could not be done in much less than a million years. To the surprise of all concerned, however, the Stanford scheme sprang a leak this year, putting mathematics known as complexity theory. Shamir was at M.I.T. in the late '70s as an associate professor of mathematics, and in fact helped write the M.I.T. code that competes head-on with Stanford's. Last spring, back in his spartan, second-floor office in the Weizmann Institute of Science in Rehovot, the lean, blue-jeaned mathematician settled the old wager: he found a way to unravel the original Stanford system. The code Shamir broke after four years of hard work was no Buck Rogers-Dick Tracy cipher. It was a charter member, along with

Figure 4.1 *Time* story

N.M. scientist cracks code

By Edward Dolnick
Globe Staff

A New Mexico computer scientist has broken a purportedly unbreakable code, foiling one of the most sophisticated schemes yet devised for keeping vital information safe from thieves and snoops.

For cracking the code, Ernest F. Brickell of the Sandia National Laboratories won a $1000 reward put up by one of the cipher's coinventors, mathematician Ralph Merkle.

"The exhilaration came when I found Merkle's check in the mail," Brickell said in a phone interview yesterday. Bets are common among the mathematicians who specialize in code-breaking, but $100 is a more typical prize than $1000, and the larger sum reflected Merkle's confidence in his cipher.

Long the province of spies and writers of thrillers, encryption—scrambling information so that outsiders cannot tap it—has lately become important in the everyday world as well.

The goal is safeguarding the growing amount of information stored in computers and protecting the flow of data transmitted by telephone or satellite. Every hour of every day, for example, an average of $13 billion is electronically transferred between banks.

Brickell, 31, worked two years on the problem known as the "knapsack" system. It is given that name, Brickell explained, because breaking the code amounts to answering a question about dividing up the contents of an enormous knapsack so they fit exactly into several smaller sacks.

Mathematically, the problem is finding which of a list of numbers add to a particular sum. For small numbers, the question can be easy: Which numbers in this list—1, 3, 5, 8—add up to 12?

Realistic examples are harder. Breaking the knapsack cipher, Brickell said, required finding which of 100 numbers, each 60 digits long, could be added together to yield a particular 62-digit sum.

Brickell said that his code-breaking method can crack the knapsack problem in under two hours on a Cray computer, one of the most powerful machines in the world.

"You can rent Cray time at about $1000 an hour," Brickell said, "so you could break the system for $2000—and you're talking about things like Electronic Funds Transfers." But even if it took $1 million to break the code it would be worth it, so great are the potential rewards.

The code that Brickell cracked, though not in commercial use, had been widely hailed as one of the most promising yet devised. It is one of a family called "public key" ciphers, invented independently in 1976 by mathematicians at MIT and Stanford.

A less powerful version of the knapsack code was cracked in 1982 (costing Merkle $100). Another public key cipher, called RSA after the initials of its inventors, remains unbroken.

Hailed as a revolution by some historians, public key ciphers appear to have crucial advantages over conventional ciphers. Those conventional codes, though complicated in practice, are straightforward in concept.

The standard approach is akin to that in the familiar a = 1, b = 2, c = 3 codes: Messages are encrypted according to some secret "recipe" or key, and then scrambled. The message can later be decoded by going through the same procedure in reverse.

But many specialists say that such systems have crucial weaknesses. The greatest one is the difficulty of getting the secret key into all the right hands, and keeping it out of all the wrong ones.

Public key ciphers, like the knapsack system, offer a mathematical end-run around that problem.

Unlike conventional ciphers with their one secret code, each user of a public key cipher has two mathematically related keys. One key, which is used to encrypt messages, is published in a public directory; the other key, used for decryption, is kept secret.

The result is that anyone can send a secret message to a Mr. Smith, say, by looking up his key in the directory. But only Smith has the key needed to decode the message, and only he can read the secret message sent to him.

11. ATTACK 12. CLAMUP 13. GET AL 14. I QUIT

In the next four problems, suppose the public key is $n = 1189$, $e = 3$. Convert each message to ciphertext.

15. SPY 16. RUB 17. KGB 18. CIA

In the next four problems, suppose the public key is $n = 1219$, $e = 7$. Convert each message to ciphertext.

19. PIG 20. COW 21. RAM 22. EWE

In the next four problems, compute b and d.

23. $p = 7, q = 11, e = 7$
24. $p = 13, q = 17, e = 5$
25. $p = 19, q = 23, e = 29$
26. $p = 31, q = 37, e = 41$

B

In the next four problems, decipher, given that $p = 23$, $q = 47$, and $e = 13$.

27. 1006, 961 28. 228, 714 29. 684, 950 30. 493, 229

In the next four problems, decipher, given that $p = 31$, $q = 37$, and $e = 11$.

31. 29, 1100 32. 382, 849 33. 324, 653 34. 454, 575

In the next four problems, apply the Euclidean algorithm to a and b, and let r_n be the last nonzero remainder. Use Problem 41 to compute integers $x_1, x_2, \ldots, x_n$ and check that $ax_n \equiv r_n \pmod{b}$.

35. $a = 217, b = 341$
36. $a = 117, b = 249$
37. $a = 143, b = 451$
38. $a = 165, b = 465$

C

Decipher each of the following two substitution messages, and determine the key.

39. RAU IBPQR QRUL RK WBQFKH BQ GUCPJBJS RAU LPKLUP JCHUQ KI RABJSQ
40. D CFVA MDQRLVAPAM F SPUIY PAJFPHFEIA NPLLT WCDRC SCDQ JFPBDK DQ SLL QJFII SL RLKSFDK
41. Suppose that a and b are positive integers. Let the Euclidean algorithm be applied to a and b, with the notation as in Theorem 1.7. Define $x_{-1} = 0$, $x_0 = 1$, and $x_i = x_{i-2} - q_i x_{i-1}$ for $i = 1, 2, \ldots, n$. Prove by induction II that $ax_i \equiv r_i \pmod{b}$ for $i = 1, 2, \ldots, n$.

Joseph Louis Lagrange

1736–1813

Lagrange, born in Turin, Italy, was of French and Italian ancestry. He became interested in mathematics as a youth, studying it on his own with such success that he was appointed professor of geometry at the Royal Artillery School in Turin at the age of 16. Thus began a long career establishing him as one of the two greatest mathematicians of the eighteenth century. (Euler was the other.)

Lagrange's modesty and nondogmatic manner enabled him to get along with diverse personalities in a contentious period of history. In 1766 he took the place of Euler, who was always generous in his praise of Lagrange's work, in the Berlin court of Frederick the Great of Prussia. When Frederick died 20 years later, Lagrange went to Paris at the invitation of Louis XVI as a member of the French Academy. Although he became a special favorite of Marie Antoinette, during the French Revolution he stayed in Paris and was given a pension and employment by the revolutionary regime, including the presidency of the commission that introduced the metric system. Later, Napoleon enjoyed conversations with him.

Lagrange is mainly known as an analyst. He made an attempt to put calculus on a rigorous foundation, which, while unsuccessful, set goals that were achieved by later mathematicians. His is the notation $f', f'', \dots$ for derivatives of a function. In algebra he is given credit for the theorem that says the number of elements in a subgroup of a finite group always divides the number of elements in the group.

Although Lagrange thought of number theory mainly as a diversion, he proved many important arithmetic results. He showed that every positive integer can be written as the sum of four squares, a result Euler worked on but failed to prove. To Lagrange is due also the first proof of Wilson's theorem, and the theorem that a congruence of degree n with prime modulus cannot have more than n solutions.

JOSEPH LOUIS LAGRANGE

Chapter 5

Congruences of Higher Degree

Just as in algebra one goes from the study of linear equations to those involving higher powers of the unknown, so also we proceed from the study of linear congruences to those of higher degree. Complications arise, however, as we will see.

5.1 Polynomial Congruences

We will consider polynomial congruences in a single unknown, for example,

$$5x^4 + 17x^3 - 3x + 2 \equiv 2x^3 - 7 \pmod{18}.$$

In fact, we will consider only congruences that can be put in the form

$$F(x) \equiv 0 \pmod{m}, \tag{5.1}$$

where $F(x)$ is a polynomial in x with integer coefficients. Our example could be put in this form by shifting everything to the left of the congruence sign, so that

$$F(x) = 5x^4 + 15x^3 - 3x + 9.$$

By Theorem 1.16 if $a \equiv a' \pmod{m}$, then $ax^k \equiv a'x^k \pmod{m}$ for all integers x. Thus coefficients of $F(x)$ may be replaced by congruent coefficients without changing the set of solutions of (5.1). For example,

$$5x^4 + 15x^3 - 3x + 9 \equiv 0 \pmod{9}$$

is equivalent to

$$5x^4 + 6x^3 - 3x \equiv 0 \pmod{9}.$$

Any term of $F(x)$ having a coefficient divisible by the modulus may be dropped completely, since it is congruent to 0 no matter what x is. These considerations motivate the following definition.

DEFINITION. degree of a congruence

If $F(x)$ is a polynomial in x with integral coefficients, then by the **degree** of the congruence

$$F(x) \equiv 0 \pmod{m}$$

we mean the exponent of the highest power of x in $F(x)$ whose coefficient is not divisible by m.

Examples. Let $F(x) = 6x^3 + x^2 + 8$. Then the degree of

$$F(x) \equiv 0 \pmod{m}$$

is 3 if $m = 5$, but if $m = 3$ then the degree is only 2.

For $m = 5$, we can find a complete solution by testing the values $x = 0, 1, 2, 3, 4$ as follows.

x	0	1	2	3	4
$F(x)$	8	15	60	179	408

Thus a complete solution is $x = 1, 2$. (Actually it would be easier to test the values $x = -2, -1, 0, 1, 2$.) Another complete solution is $x = 42, 16$ (since $42 \equiv 2$ and $16 \equiv 1 \pmod{5}$.) ♢

Reducing a Congruence to Prime Power Moduli

Consider the congruence

$$F(x) = 6x^3 + x^2 + 8 \equiv 0 \pmod{20}.$$

Solving this by trial-and-error would involve evaluating $F(x)$ for 20 integers, a tedious task. There is a way to reduce the work, however. Recall that Theorem 4.3 says that if $(b_1, b_2) = 1$, then

$$z \equiv z' \pmod{b_1 b_2} \text{ if and only if } z \equiv z' \pmod{b_1} \text{ and } z \equiv z' \pmod{b_2}.$$

Applying this theorem with $z = F(x)$, $z' = 0$, $b_1 = 5$, and $b_2 = 4$, we see that

$$F(x) \equiv 0 \pmod{20}$$

is equivalent to

$$F(x) \equiv 0 \pmod{5} \quad \text{and} \quad F(x) \equiv 0 \pmod{4}.$$

We solved the congruence with modulus 5 in the last example, finding that $x = 1, 2$ was a complete solution. From the table of values of $F(x)$ above, we see that $x = 0, 2$ is a complete solution for the modulus 4. Then x is a solution to the original congruence if and only if

$$x \equiv 1 \text{ or } 2 \pmod 5 \quad \text{and} \quad x \equiv 0 \text{ or } 2 \pmod 4. \tag{5.2}$$

Solving these simultaneous congruences can be done by the Chinese remainder theorem, discussed in Section 4.2. Each pair of solutions modulo 4 and 5 generates a unique solution modulo 20, so we get four (2 times 2) solutions in all. The reader should review the Chinese remainder theorem and confirm that x satisfies (5.2) if and only if $x \equiv 2,\ 6,\ 12,$ or $16 \pmod{20}$. Thus this is a complete solution to the original congruence.

Notice that in order to apply Theorem 4.3 and the Chinese remainder theorem the smaller moduli employed must be relatively prime. Thus factoring 20 into 4 times 5 worked, but 2 times 10 would not have, since 2 and 10 are not relatively prime.

This method may be used to simplify the solution of

$$F(x) \equiv 0 \pmod m$$

whenever m can be factored into relatively prime smaller factors, that is, whenever more than one prime divides m. More than two factors may be used. For example, $F(x) \equiv 0 \pmod{360}$ may be replaced by

$$F(x) \equiv 0 \pmod 9 \quad \text{and} \quad F(x) \equiv 0 \pmod{40}.$$

But the latter congruence is equivalent to

$$F(x) \equiv 0 \pmod 8 \quad \text{and} \quad F(x) \equiv 0 \pmod 5.$$

The smallest pieces the modulus can be broken into are the prime powers in its factorization. This method is summarized in the following theorem. The details of the proof (a version of Theorem 4.3 for more than two moduli is needed, for example) are left for the exercises.

Theorem 5.1. *Consider the congruence*

$$F(x) \equiv 0 \pmod m,$$

where F is a polynomial with integer coefficients and m is a positive integer. Let $m = p_1^{n_1} p_2^{n_2} \ldots p_r^{n_r}$, where $p_1, p_2, \ldots, p_r$ are distinct primes. Suppose a complete solution to

$$F(x) \equiv 0 \pmod{p_i^{n_i}}$$

has k_i elements, $i = 1, 2, \ldots, r$. Then a complete solution to the original congruence has $k_1 k_2 \cdots k_r$ elements. In fact, if for each i, $i = 1, 2, \ldots, r$, x_i is a solution to

$$F(x) \equiv 0 \pmod{p_i^{n_i}},$$

then any x such that

$$x \equiv x_i \pmod{p_i^{n_i}}, \quad i = 1, 2, \dots, r$$

satisfies the original congruence, and a complete solution to it may be constructed this way by allowing the x_i to run through complete solutions to the congruences with prime power moduli.

Example. Solve $3x^2 - 20x + 25 \equiv 0 \pmod{84}$.

The prime power factors of 84 are 4, 3, and 7. By testing values in complete residue systems we find that a complete solution to

$$3x^2 + 1 \equiv 0 \pmod 4$$

is $x = 1, 3$; a complete solution to

$$x + 1 \equiv 0 \pmod 3$$

is $x = -1$; and a complete solution to

$$3x^2 + x + 4 \equiv 0 \pmod 7$$

is $x = -3, -2$. (Note that the congruences have been simplified by reducing coefficients, depending on the modulus.)

Now we use the Chinese remainder theorem to solve the simultaneous congruences

$$\begin{aligned} x &\equiv 1 \text{ or } 3 \pmod 4 \\ x &\equiv -1 \pmod 3 \\ x &\equiv -3 \text{ or } -2 \pmod 7, \end{aligned}$$

which involves solving

$$\begin{aligned} 21x_1 &\equiv 1 \pmod 4, \\ 28x_2 &\equiv 1 \pmod 3, \\ 12x_3 &\equiv 1 \pmod 7. \end{aligned}$$

Solutions are $x_1 = 1$, $x_2 = 1$, and $x_3 = 3$. Then the simultaneous solutions are

$$\begin{aligned} x &= 21(1)(1 \text{ or } 3) + 28(1)(-1) + 12(3)(-3 \text{ or } -2) \\ &= -115, -79, -37, -73. \end{aligned}$$

This is a complete solution to the original congruence. The least complete solution is $x = 5, 11, 47, 53$. ♢

Example. Solve $x^2 + 1 \equiv 0 \pmod{35}$.

By trial we find that of the two congruences

$$x^2 + 1 \equiv 0 \pmod 5 \quad \text{and} \quad x^2 + 1 \equiv 0 \pmod 7,$$

$x = -2, 2$ is a complete solution to the first, but the second congruence has no solutions. Thus the original congruence has no solutions either. This illustrates that in Theorem 5.1 some of the numbers k_i may be zero. ♢

Congruences with Modulus p^2

The above method does not help with the congruence

$$2x^3 - 3x + 6 \equiv 0 \pmod{25}, \tag{5.3}$$

since only one prime divides the modulus. It is possible to avoid testing 25 values of x in it, however. The method we will illustrate depends on the fact that by part (6) of Theorem 1.16 any solution to (5.3) also satisfies

$$2x^3 - 3x + 6 \equiv 0 \pmod{5}. \tag{5.4}$$

By testing $x = -2, -1, 0, 1, 2$ in (5.4) we see it has a complete solution $x = 1$.

Thus instead of trying all of 0, 1, ... , 24 in (5.3) we need only try values of $x \equiv 1 \pmod{5}$, namely, $x =$ 1, 6, 11, 16, and 21.

Even these computations may be avoided. We want to test solutions of the form $x = 1 + 5y$ in (5.3), where $y = 0, 1, 2, 3$, or 4. Substituting this expression in (5.3) gives

$$2(1 + 5y)^3 - 3(1 + 5y) + 6 \equiv 0 \pmod{25}.$$

Simplifying this algebraically and dropping those terms with coefficients divisible by 25 yields

$$15y + 5 \equiv 0 \pmod{25}.$$

By Theorem 1.19 (with $a = 5$ and $b = 25$) this is equivalent to

$$3y + 1 \equiv 0 \pmod{5}.$$

Since we are interested in y ranging from 0 to 4, we want a complete solution. One is easily seen to be $y = 3$. We see that a complete solution to (5.3) is $x = 1 + 5(3) = 16$.

The method of this example works in general. That is, given a congruence

$$F(x) \equiv 0 \pmod{p^2}, \tag{5.5}$$

where p is prime, we first solve

$$F(x) \equiv 0 \pmod{p}. \tag{5.6}$$

Now if x' is any solution to (5.6), we look for solutions to (5.5) of the form $x = x' + py$, where y is in some complete residue system modulo p. Substituting this expression in (5.5) leads to a *linear* congruence

$$ay + b \equiv 0 \pmod{p} \tag{5.7}$$

in y, to which a complete solution is desired. The reason that (5.7) turns out to be linear is that when $x' + py$ is raised to a power, most of the terms involving y drop out because thay have a coefficient divisible by p^2.

Note that if (5.5) has more than one solution x', the succeeding process must be applied to each of them. This technique will be generalized and investigated in detail in the next section.

Problems for Section 5.1.

A *In the first 10 problems, find the least complete solution by testing.*

1. $x^3 - x^2 + 3x + 1 \equiv 0 \pmod{7}$
2. $x^3 - x^2 + 3x + 1 \equiv 0 \pmod{5}$
3. $x^2 - 3x + 1001 \equiv 0 \pmod{3}$
4. $x^2 - 3x + 1001 \equiv 0 \pmod{5}$
5. $x^2 - 3x + 1001 \equiv 0 \pmod{7}$
6. $x^4 - x^3 + 5x + 1 \equiv 0 \pmod{3}$
7. $x^4 - x^3 + 5x + 1 \equiv 0 \pmod{11}$
8. $x^4 - x^3 + 5x + 1 \equiv 0 \pmod{4}$
9. $x^7 - x \equiv 0 \pmod{7}$
10. $x^7 - x \equiv 0 \pmod{5}$

In the next four problems, find all solutions x with $|x| < 10$.

11. $x^3 + x + 3 \equiv 0 \pmod{7}$
12. $x^3 + x + 3 \equiv 0 \pmod{13}$
13. $x^3 + x + 3 \equiv 0 \pmod{19}$
14. $x^3 + x + 3 \equiv 0 \pmod{3}$

In the next 10 problems, break the modulus into prime powers to find the least complete solution.

15. $x^3 - x^2 + 3x + 1 \equiv 0 \pmod{35}$
16. $x^2 - 3x + 1001 \equiv 0 \pmod{15}$
17. $x^2 - 3x + 1001 \equiv 0 \pmod{10}$
18. $x^2 - 3x + 1001 \equiv 0 \pmod{21}$
19. $x^4 - x^3 + 5x + 1 \equiv 0 \pmod{33}$
20. $x^4 - x^3 + 5x + 1 \equiv 0 \pmod{66}$
21. $x^2 + x + 1 \equiv 0 \pmod{21}$
22. $x^2 + x + 1 \equiv 0 \pmod{91}$
23. $x^2 + x + 1 \equiv 0 \pmod{273}$
24. $x^2 + x + 1 \equiv 0 \pmod{195}$

B

In the next four problems, a congruence is given with modulus p^2, p a prime. A solution x' to the corresponding congruence with modulus p is also given. Substitute $x = x' + py$ into the congruence and find a linear congruence with modulus p that y must satisfy as in the example at the end of this section. Find any corresponding solutions x, $0 \leq x < p^2$.

25. $x^3 + 8 \equiv 0 \pmod{9}$, $x' = 1$
26. $x^2 + x + 1 \equiv 0 \pmod{49}$, $x' = 4$
27. $x^3 + 8 \equiv 0 \pmod{25}$, $x' = 3$
28. $x^2 + x + 1 \equiv 0 \pmod{49}$, $x' = 2$

In the next four problems, use the method at the end of this section to find the least complete solution.

29. $x^2 - x + 2 \equiv 0 \pmod{25}$
30. $x^2 - x + 2 \equiv 0 \pmod{49}$
31. $x^2 + x + 2 \equiv 0 \pmod{121}$
32. $x^3 + 8 \equiv 0 \pmod{49}$

C

In the next four problems, let k and k' be the number of elements in complete solutions to $F(x) \equiv 0 \pmod{3}$ and $F(x) \equiv 0 \pmod{9}$, respectively. Construct an integral polynomial F so that k and k' are as given.

33. $k = 3$, $k' = 0$
34. $k = 1$, $k' = 3$
35. $k = 1$, $k' = 1$
36. $k = 2$, $k' = 4$
37. Use induction on n to prove that if $m = m_1 m_2 \cdots m_n$ and $m_1, m_2, \ldots, m_n$ are positive integers, relatively prime in pairs, then $z \equiv z' \pmod{m}$ if and only if $z \equiv z' \pmod{m_i}$ for $i = 1, 2, \ldots, n$.
38. Prove Theorem 5.1

5.2 Congruences with Prime Power Moduli

Generalizing the Method of the Last Section

In the last section, we saw how solving any polynomial congruence can be reduced, with the help of the Chinese remainder theorem, to solving congruences with moduli powers of primes; and at the end of the section a method of solving congruences with modulus p^2 was illustrated. We will generalize the latter method to solve congruences with modulus p^n, where p is prime.

The general plan, given a congruence

$$F(n) \equiv 0 \pmod{p^n},$$

will be to find first a complete solution to

$$F(n) \equiv 0 \pmod{p},$$

then use this to solve

$$F(n) \equiv 0 \pmod{p^2},$$

and continue this way until we have a complete solution to the original congruence. What we need is a method for going from modulus p^k to modulus p^{k+1}.

Consider the congruences

$$F(x) \equiv 0 \pmod{p^k}, \tag{5.8}$$

and

$$F(x) \equiv 0 \pmod{p^{k+1}}. \tag{5.9}$$

By the last part of Theorem 1.16 any solution of (5.9) is also a solution of (5.8), so if $x_k, x'_k, x''_k, \ldots$ is a complete solution of (5.8), then any solution of (5.9) will be congruent to one of these numbers $\pmod{p^k}$.

Thus we can restrict our attention to solutions of (5.9) of the form

$$x_k + p^k y, x'_k + p^k y, \ldots,$$

where y is an integer. Furthermore, we can assume that y is in some complete residue system $\pmod{p}$, since if $y \equiv y' \pmod{p}$, then it can be checked that

$$x_k + p^k y \equiv x_k + p^k y' \pmod{p^{k+1}},$$

and so these numbers could not both appear in a complete solution to (5.9).

Let us consider a particular solution x_k of (5.8) and see what solutions of the form $x_k + p^k y$ of (5.9) it might generate. We wish to find out which y satisfy

$$F(x_k + p^k y) \equiv 0 \pmod{p^{k+1}}. \tag{5.10}$$

Let us assume

$$F(x) = a_r x^r + a_{r-1}x^{r-1} + \cdots + a_0.$$

Then a typical term on the left side of (5.10) will be

$$a_j(x_k + p^k y)^j.$$

If this is multiplied out, an unpleasant expression of $j+1$ terms will result; but fortunately it turns out that all but the first two will be divisible by p^{k+1}, and so may be dropped from the congruence (5.10).

Since the complexity of the expresssion involved may obscure what is going on, let us simplify the notation. We claim that

$$(u + mv)^j = u^j + ju^{j-1}mv + (\text{a multiple of } m^2)$$

for any integers u, m, and v, and any positive integer j. Although the binomial theorem could be invoked to prove this, all that is really needed is to notice that the expression on the left is $u + mv$ multiplied by itself j times, that is

$$\underbrace{(u + mv)(u + mv)\cdots(u + mv)}_{j \text{ factors}}.$$

Forming a term by choosing all us gives u^j. Choosing one factor mv and the rest us can be done in j ways, and so gives a term $ju^{j-1}mv$. All other terms contain the factor mv at least twice, and so are multiples of m^2.

In particular, if we take $u = x_k$, $m = p^k$, and $v = y$, we see that

$$a_j(x_k + p^k y)^j = a_j(x_k^j + jx_k^{j-1}p^k y + \text{a multiple of } p^{2k}).$$

Since it is easy to see that $2k \geq k+1$ for k any positive integer, this means that in the congruence (5.10) the term in question may be replaced by

$$a_j x_k^j + a_j j x_k^{j-1} p^k y.$$

By collecting the terms involving y we see that (5.10) may be written in the form

$$F(x_k) + F'(x_k)p^k y \equiv 0 \pmod{p^{k+1}},$$

where $F'(x)$ is the polynomial

$$a_r r x^{r-1} + a_{r-1}(r-1)x^{r-2} + \cdots + a_1.$$

Calculus students should recognize that this is just the derivative of $F(x)$. No calculus is really needed here, however, since it is easy enough to define $F'(x)$ in a formal way for any polynomial $F(x)$.

DEFINITION. derivative of a polynomial

Given a polynomial $F(x) = a_r x^r + \cdots + a_0$, we define its (formal) **derivative** to be the polynomial

$$F'(x) = a_r r x^{r-1} + a_{r-1}(r-1)x^{r-2} + \cdots + a_1.$$

For example, the derivative of

$$3x^5 - 4x^3 + 2x^2 + 9x + 14$$

is

$$15x^4 - 12x^2 + 4x + 9.$$

A Linear Congruence for y

We saw above that if x_k is a solution to (5.8), then $x_k + p^k y$ is a solution to (5.9) exactly when

$$F(x_k) + F'(x_k)p^k y \equiv 0 \pmod{p^{k+1}}.$$

By Theorem 1.19 this congruence is equivalent to

$$\frac{F(x_k)}{p^k} + F'(x_k)y \equiv 0 \pmod{p},$$

or

$$F'(x_k)y \equiv \frac{-F(x_k)}{p^k} \pmod{p}.$$

(Note that the term on the right is an integer by the assumption that x_k satisfies (5.8).) Thus finding y involves only solving a linear congruence with modulus p.

Before summarizing this method in a theorem we illustrate it.

Example. Solve $x^2 + x + 3 \equiv 0 \pmod{27}$.

We start by solving the congruence

$$x^2 + x + 3 \equiv 0 \pmod{3}$$

by testing values of x in a complete residue system modulo 3, and find a complete solution is $x_1 = -1, 0$.

Solutions to the congruence with modulus 9 will be of the form $x_1 + 3y$, where y satisfies

$$F'(x_1)y \equiv \frac{-F(x_1)}{3} \pmod{3}.$$

Note that $F'(x) = 2x + 1$. For $x_1 = -1$, this congruence is

$$(-1)y \equiv \frac{-3}{3} \pmod{3}.$$

This has the complete solution $y = 1$, giving $x_2 = -1 + 3(1) = 2$.

If we take $x_1 = 0$, then the congruence is

$$(1)y \equiv \frac{-3}{3} \pmod{3},$$

which has the complete solution $y = -1$, giving $x_2 = 0 + 3(-1) = -3$.

Now we look for solutions to the original congruence with modulus 27 of the form $x_3 = x_2 + 9y$. Here y must satisfy

$$F'(x_2)y \equiv \frac{-F(x_2)}{9} \pmod{3}.$$

For $x_2 = 2$, this congruence is

$$5y \equiv \frac{-9}{9} \pmod{3},$$

which has the complete solution $y = 1$; and so $x_3 = 2 + 9(1) = 11$.

Using $x_2 = -3$ gives the congruence

$$-5y \equiv \frac{-9}{9} \pmod{3},$$

which has the complete solution $y = -1$. Thus here $x_3 = -3 + 9(-1) = -12$. A complete solution to the original congruence is $x = 11, -12$. The least complete solution is $x = 11, 15$. ◇

Theorem 5.2. *Suppose $x_k, x'_k, x''_k, \ldots$ is a complete solution to the congruence $F(x) \equiv 0 \pmod{p^k}$, where $F(x)$ is an integral polynomial and p is a prime. Then all solutions to $F(x) \equiv 0 \pmod{p^{k+1}}$ in some complete residue system modulo p^{k+1} congruent to $x_k \pmod{p^k}$ are given by $x_{k+1} = x_k + p^k y$, where y runs through any complete solution to*

$$F'(x_k)y \equiv \frac{-F(x_k)}{p^k} \pmod{p}.$$

Applying this also to $x'_k, x''_k, \ldots$ in turn yields a complete solution to $F(x) \equiv 0 \pmod{p^{k+1}}$.

Notice that according to Theorem 4.2 the congruence $ax \equiv c \pmod{b}$ has a solution if and only if (a, b) divides c, and if solutions exist, then a complete solution has (a, b) elements. In the case of the congruence defining y, we have the modulus b is p, and so (a, b) equals 1 or p. Thus each solution x_k generates 0, 1, or p solutions x_{k+1} at the next level.

Example. Solve $x^3 + x^2 + 23 \equiv 0 \pmod{125}$.

We start with the congruence $F(x) \equiv 0 \pmod{5}$, where $F(x) = x^3 + x^2 + 23$. The following table shows that a complete solution is $x_1 = 1, 2$.

x_1	-2	-1	0	1	2
$F(x_1)$	19	23	23	25	35

Notice that $F'(x) = 3x^2 + 2x$. We will find solutions x_2 to $F(x) \equiv 0 \pmod{25}$ of the form $x_1 + 5y$, where y satisfies

$$F'(x_1)y \equiv \frac{-F(x_1)}{5} \pmod{5}.$$

For $x_1 = 1$, this congruence is

$$5y \equiv \frac{-25}{5},$$

and a complete solution is $y = -2, -1, 0, 1, 2$. Thus $x_2 = 1 + 5y = -9, -4, 1, 6, 11$.

For $x_1 = 2$, the congruence for y becomes

$$16y \equiv \frac{-35}{5} \pmod{5},$$

and a complete solution is $y = -2$, yielding $x_2 = 2 + 5y = -8$.

It is useful to make a table of F and F' for the various values of x_2.

x_2	-9	-4	1	6	11	-8
$F(x_2)$	-625	-25	25	275	1475	-425
$F'(x_2)$	$\equiv 5$	$\equiv 5$	$\equiv 5$	$\equiv 5$	$\equiv 5$	$\equiv 16$

Here the congruences in the last line are modulo 5. Notice that in the congruence

$$F'(x_k)y \equiv \frac{-F(x_k)}{p^k} \pmod{p}$$

all that matters about $F'(x_k)$ is what it is modulo p. Since all solutions x_k arising from a given solution x_1 at the first level will be congruent modulo p, F' need only be computed once for each of these. For example, all the solutions $x_2 = -9, -4, 1, 6, 11$ arising from $x_1 = 1$ will have $F'(x_2) \equiv 5 \pmod{5}$.

We must find new values of y satisfying

$$F'(x_2)y \equiv \frac{-F(x_2)}{25} \pmod{5}.$$

For $x_2 = -9, -4, 1, 6, 11$, we get the congruences

$$5y \equiv 25, 1, -1, -11, -59 \pmod{5}.$$

Only the first of these is solvable, with the complete solution $y = -2, -1, 0, 1, 2$, which yields $x_3 = -9 + 25y = -59, -34, -9, 16, 41$.

Taking $x_2 = -8$ leads to the congruence

$$16y \equiv 17 \pmod{5}$$

with the complete solution $y = 24$, which gives $x_3 = -8 + 25y = 42$.

We see that a complete solution to the original congruence is $x = -59, -34, -9, 16, 41, 42$. The least complete solution is $x = 16, 41, 42, 66, 91, 116$. ◇

Problems for Section 5.2.

A *In the first six problems, a polynomial $F(x)$ is given. Find the derivative $F'(x)$.*

1. $5x^2 + 3x - 7$
2. $-3x^4 - 2x^2 + 4x + 2$
3. $7x^{10} - 2x^7 + 3x$
4. $9x^8 + 2x^5 - x + 13$
5. $(x+1)^3$
6. $(x+2)^4$

In the next 12 problems, find the least complete solution.

7. $x^2 - x + 7 \equiv 0 \pmod{9}$
8. $x^2 + 3x - 50 \equiv 0 \pmod{25}$
9. $x^2 - x + 7 \equiv 0 \pmod{27}$
10. $x^2 + 3x - 50 \equiv 0 \pmod{125}$
11. $x^2 - 44 \equiv 0 \pmod{27}$
12. $x^2 - 44 \equiv 0 \pmod{81}$
13. $x^3 + x^2 + 10 \equiv 0 \pmod{32}$
14. $x^3 + x^2 + 1 \equiv 0 \pmod{27}$
15. $x^3 + x^2 + 1 \equiv 0 \pmod{81}$
16. $x^4 + x + 23 \equiv 0 \pmod{25}$
17. $x^4 + x + 23 \equiv 0 \pmod{125}$
18. $x^3 + 13 \equiv 0 \pmod{49}$

B

19. Show that if $y \equiv y' \pmod{p}$, then $x_k + p^k y \equiv x_k + p^k y' \pmod{p^{k+1}}$, as is claimed in the proof of Theorem 5.2.
20. Prove by induction on n that if m and n are positive integers, then $(u + mv)^n \equiv u^n + nu^{n-1}mv \pmod{m^2}$.

In the next six problems, find the least complete solution.

21. $x^2 - x - 12 \equiv 0 \pmod{100}$
22. $10x^2 - x^3 \equiv 9 \pmod{405}$
23. $x^2 - x - 12 \equiv 0 \pmod{1000}$
24. $x^2 + 3x - 50 \equiv 0 \pmod{275}$
25. $x^2 - x + 7 \equiv 0 \pmod{63}$
26. $x^2 - x + 7 \equiv 0 \pmod{117}$

27. Find all x, $0 \le x < 72$, such that $x^2+3x+6 \equiv 0 \pmod{8}$ and $x^3+x^2+3 \equiv 0 \pmod{9}$.
28. Suppose F is an integral polynomial, p is a prime, and S is a finite set of integers such that given $n > 0$ there exists x_n in S such that $F(x_n) \equiv 0 \pmod{p^n}$. Show there exists x_0 in S such that $F(x_0) = 0$.

C

29. Prove the **binomial theorem** by induction on n:

$$(a+b)^n = a^n + C(n,1)a^{n-1}b + C(n,2)a^{n-2}b^2 + \cdots + C(n,n-1)ab^{n-1} + b^n$$

for all positive integers n, where $C(n,r) = n!/r!(n-r)!$.

30. Since $(a+b)^n$ is clearly an integral polynomial in a and b, the binomial theorem shows that $n!/r!(n-r)!$ is an integer for $1 < r < n$. A direct proof is not so easy to come by. Criticize the following argument: The r consecutive integers $n, n-1, \ldots, n-r+1$ contain a complete residue system modulo k for $1 \le k \le r$. Thus one of them is divisible by k. Since this works for $k = 1, 2, \ldots, r$ we conclude that $r!$ divides $n(n-1)\cdots(n-r+1) = n!/(n-r)!$.
31. Show that if a and b are positive integers and $b = aq + r$, $0 \le r < a$, then exactly q of the integers $1, 2, \ldots, b$ are divisible by a.
32. Suppose p is a prime, b is a positive integer, and $b = q_i p^i + r_i$, $0 \le r_i < p^i$ for $i = 1, 2, \ldots$. Let p^t divide $b!$ exactly. Prove that $t = q_1 + q_2 + \cdots$. (See Section 2.3 for the definition of dividing exactly.)

33. Suppose a, b, and b' are positive integers, with $b+b'=b''$. Let $b=aq+r$, $b'=aq'+r'$, and $b''=aq''+r''$, with $0 \le r, r', r'' < a$. Show that $q+q' \le q''$.
34. Suppose p is prime and b and b' are positive integers, with $b+b'=b''$. Suppose p^t and $p^{t'}$ exactly divide $b!$ and $b'!$, respectively. Show that $p^{t+t'}$ divides $b''!$.
35. Show that if n is a positive integer and $0 \le r \le n$, then $r!(n-r)!$ divides $n!$.
36. Show that $3^{48} \parallel 100!$.
37. Solve $2^x \parallel 100!$.
38. Solve $7^y \parallel 462!$.
39. Solve $6^z \mid 45!$, $6^{z+1} \nmid 45!$.
40. Prove or disprove: If p and q are primes, $p < q$, and if $q^k \mid n!$, then $p^k \mid n!$.
41. Prove that if n and r are nonnegative integers, then $r!$ divides $(n+1)(n+2)\cdots(n+r)$.

In the next three problems, assume that F is an integral polynomial and p is a prime.

42. Prove or disprove: If $F(x) \equiv 0 \pmod{p^n}$ is solvable for all positive integers n, then there exists an x_0 such that $F(x_0)=0$.
43. Suppose $p \mid F(x_0)$ but $p \nmid F'(x_0)$. Show that $F(x) \equiv 0 \pmod{p^n}$ has a solution for all positive integers n.
44. Let h be the number of elements in a complete solution to $F(x) \equiv 0 \pmod{p^2}$ and suppose $p \nmid h$. Show that $F(x) \equiv 0 \pmod{p^n}$ has a solution for all positive integers n.
45. For what integers h does there exist an integral polynomial F such that a complete solution to $F(x) \equiv 0 \pmod{9}$ has exactly h elements?

5.3 Quadratic Residues

Congruences of Degree Two

In the previous two sections, we have seen how to reduce the solution of any polynomial congruence to that of congruences with prime moduli. For such congruences, however, our only technique is to test the elements of a complete residue system.

A reasonable way to start the analysis of such congruences would be to restrict their degree. Since linear congruences were completely covered in Section 4.1, we consider congruences of degree two, say

$$ax^2 + bx + c \equiv 0 \pmod{p}, \tag{5.11}$$

where p is prime.

The first thing that comes to mind on looking at (5.11) is the quadratic formula. This says that the solutions of the *equation* $ax^2 + bx + c = 0$ are given by

$$x = \frac{-b \pm \sqrt{b^2 - 4ac}}{2a}.$$

If we are interested only in real solutions there will be 0, 1, or 2 of them according as $b^2 - 4ac$ is negative, zero, or positive.

In the congruence (5.11), we are looking for integral solutions x. Of course, if $b^2 - 4ac$ is not a perfect square, then the radical will not be an integer (it may not even be real). Perhaps it would be sufficient for $b^2 - 4ac$ to be *congruent* to a square, say y^2, modulo p. Then we might take

$$x = \frac{-b + y}{2a}.$$

A problem still remains, namely, the division by $2a$, which may not produce an integer. Why not just require x to be a solution to

$$2ax \equiv -b + y \pmod{p}?$$

By Theorem 4.2 this will exist whenever $(2a, p) = 1$. But we can assume that p does not divide a, since otherwise (5.11) is not really a congruence of degree 2. Also if p divides 2, then $p = 2$. We should be so lucky; solving congruences with modulus 2 is no great problem.

Our musings have led us to the notion that if $y^2 \equiv b^2 - 4ac \pmod{p}$, and if x is a solution to $2ax \equiv -b + y \pmod{p}$, then x is a solution to the congruence (5.11). We might even dream that every solution to (5.11) could be generated this way.

The strange thing is that it works! We have just seen an example of a type of reasoning, compounded of intuition, guesswork, and wishful thinking, that pervades mathematical creation yet almost never gets into print. Although sometimes completely fruitless, such uncritical thinking aften carries one close enough to the truth to direct the application of more rigorous arguments, just as an illegal wiretap, while not admissible in court, may lead the police to construct a legitimate case.

Theorem 5.3. *If a, b, and c are integers and p is an odd prime not dividing a, then the solutions of the congruence*

$$ax^2 + bx + c \equiv 0 \pmod{p}$$

are given by the solutions to $2ax \equiv -b + y \pmod{p}$, where y runs through a complete solution to $y^2 \equiv b^2 - 4ac$. A complete solution has 1 element if p divides $b^2 - 4ac$, otherwise 0 or 2 elements.

Proof. To prove the first sentence we must show two things:

1. If $y^2 \equiv b^2 - 4ac$ and $2ax \equiv -b + y \pmod{p}$, then x is a solution to the original congruence.

2. Every solution to the original congruence is generated this way.

As with the proof of the quadratic formula proving (1) merely involves substitution into the original congruence, and will be left for the exercises at the end of this section.

To prove (2) we also copy the corresponding part of the proof of the quadratic formula, which involves completing the square. Let us assume x satisfies the original congruence. We multiply by $4a$ to be sure that everything remains integral when we complete the square, getting

$$4a^2x^2 + 4abx + 4ac \equiv 0 \pmod{p},$$

or

$$(2ax + b)^2 - b^2 + 4ac \equiv 0 \pmod{p}.$$

If we now define y to be $2ax + b$ then we have from the last congruence that $y^2 \equiv b^2 - 4ac \pmod{p}$, and also that $2ax \equiv -b + y \pmod{p}$ from the definition of y.

Notice that by Theorem 4.5 if y generates x and y' generates x' by this method, then $x \equiv x' \pmod{p}$ if and only if $y \equiv y' \pmod{p}$. Thus to count the elements in a complete solution to the original congruence it suffices to count a complete solution to $y^2 \equiv b^2 - 4ac \pmod{p}$. Now if y is a solution and y' is any other, then $y^2 \equiv y'^2 \pmod{p}$, so p divides $y^2 - y'^2 = (y - y')(y + y')$. By Theorem 2.4 this implies $y' \equiv y$ or $-y \pmod{p}$. There are two solutions unless $y \equiv -y \pmod{p}$, which says p divides $2y$. Since p is odd the latter happens only when p divides y (and so y^2), in which case there is only one solution. But $y^2 \equiv b^2 - 4ac \pmod{p}$, which proves the last sentence of the theorem. □

Example. Solve $x^2 + 6x + 1 \equiv 0 \pmod{31}$.

Here $a = 1$, $b = 6$, and $c = 1$. The obvious solutions to

$$y^2 \equiv b^2 - 4ac = 32 \equiv 1 \pmod{31}$$

are $y = \pm 1$. Then we must solve

$$2ax \equiv -b + y \quad \text{or} \quad 2x \equiv -6 \pm 1 = -5 \text{ or } -7 \pmod{31}.$$

Solutions are $x = 13, 12$; and this is a complete solution to the original congruence. ♢

Quadratic Residues

Clearly it is of interest to know for which numbers m the congruence $y^2 \equiv m \pmod{p}$ is solvable. If p divides m, as we have seen, $y = 0$ is a complete solution, but if $p \nmid m$ whether a solution exists or not may not be obvious.

DEFINITION. quadratic residue, quadratic nonresidue

Let p be prime and suppose $p \nmid a$. We call a a **quadratic residue modulo** p in case there exists an integer y such that $y^2 \equiv a \pmod{p}$. If no such y exists then we call a a **quadratic nonresidue modulo** p.

Examples. The integers 1, 4, and 12 are quadratic residues (mod 13) (note that $5^2 \equiv 12 \pmod{13}$), while 2 is a quadratic nonresidue (mod 5) since $y^2 \equiv 2 \pmod{5}$ has no solution. ♢

Since if $a \equiv a' \pmod{p}$, then the congruences

$$y^2 \equiv a \pmod{p} \quad \text{and} \quad y^2 \equiv a' \pmod{p}$$

are equivalent, the quadratic residues make whole congruence classes. The argument at the end of the proof of the previous theorem shows that if p is an odd prime, if $p \nmid a$, and if $y^2 \equiv a \pmod{p}$ is solvable, then there are exactly two elements in a complete solution, the other being congruent to $-y$. Thus if we compute

$$1^2, 2^2, 3^2, \ldots, (p-1)^2,$$

we hit each congruence class (mod p) containing quadratic residues exactly twice, first with some k, $1 \leq k \leq (p-1)/2$, and the second time with $p - k$, since $(p-k)^2 \equiv k^2 \pmod{p}$. Thus we have the following theorem, which at least cuts down by half the task of looking for solutions to $y^2 \equiv a \pmod{p}$.

Theorem 5.4. *Let p be any odd prime. Then any reduced residue system modulo p contains $(p-1)/2$ quadratic residues and $(p-1)/2$ quadratic nonresidues modulo p. One set of $(p-1)/2$ incongruent quadratic residues is*

$$1^2, 2^2, \ldots, \left(\frac{p-1}{2}\right)^2.$$

Example. We compute the least residues (mod 11) of k^2, $k = 1, 2, \ldots, 10$ in the following table.

k	1	2	3	4	5	6	7	8	9	10
$k^2 \equiv$	1	4	9	5	3	3	5	9	4	1

Note that each quadratic residue appears once between $k = 1$ and $5 = (11 - 1)/2$. ◇

Example. Solve $2x^2 + 3x + 5 \equiv 0 \pmod{23}$.

Here $b^2 - 4ac = 9 - 4(2)(5) = -31 \equiv 15 \pmod{23}$. We compute the least residues (mod 23) of the squares of the integers from 1 to $(23-1)/2 = 11$ to get the following table.

k	1	2	3	4	5	6	7	8	9	10	11
$k^2 \equiv$	1	4	9	16	2	13	3	18	12	8	6

Letting k run from 12 to 22 would produce the same squares in reverse order. (Try it!) Thus 15 is a quadratic nonresidue (mod 23) and the original congruence has no solution. ◇

A Variation on Wilson's Theorem

The reader may want to review the proof of Wilson's theorem, in which it is shown that $(p-1)! \equiv -1 \pmod{p}$ by matching each of the integers k from $1, 2, \ldots, p-1$ with the integer k' from that list such that $kk' \equiv 1 \pmod{p}$. We vary this proof as follows. Suppose p is an odd prime not dividing a, and match each k between 1 and $p-1$ with a k' in that set such that $kk' \equiv a \pmod{p}$. Exactly one such k exists by Theorem 4.5.

There is a problem with the count if we could have $k = k'$, which says $k^2 \equiv a \pmod{p}$. This problem cannot arise if a is a quadratic nonresidue (mod p), so let us assume that such is the case for the time being. Then we have $(p-1)/2$ pairs kk', each with $kk' \equiv a \pmod{p}$. Thus

$$a^{(p-1)/2} \equiv 1 \cdot 2 \cdot \cdots \cdot (p-1) \equiv -1 \pmod{p},$$

where, of course, the last congruence is Wilson's theorem. We see that if a is a quadratic nonresidue (mod p), then $a^{(p-1)/2} \equiv -1 \pmod{p}$.

If a is a quadratic residue (mod p) we take a more direct approach. Let $y^2 \equiv a \pmod{p}$. Since $p \nmid a$, also $p \nmid y$. Then

$$a^{(p-1)/2} \equiv y^{p-1} \equiv 1 \pmod{p}$$

by Fermat's theorem.

Theorem 5.5 (Euler's criterion). *Let p be an odd prime not dividing the integer a. Then a is a quadratic residue or quadratic nonresidue modulo p according as*

$$a^{(p-1)/2} \equiv 1 \text{ or } -1 \pmod{p}.$$

In light of the modular exponentiation algorithm of Section 4.4, this theorem gives an efficient method of telling whether a given integer is a quadratic residue or nonresidue modulo p, although it does not give an actual y such that $y^2 \equiv a \pmod{p}$ in the case of a quadratic residue. The case $a = -1$ is especially pleasant, since powers of -1 are easily computed.

Corollary 5.6. *The number -1 is a quadratic residue or quadratic nonresidue modulo the odd prime p according as p is congruent to 1 or 3 modulo 4.*

Proof. Of course, any odd prime p must be congruent to 1 or 3 modulo 4, and it is easy to check that then $(p-1)/2$ is even or odd, respectively. □

DEFINITION. Legendre symbol

Suppose p is an odd prime not dividing the integer a. We define the symbol (a/p) to be 1 or -1 according as a is a quadratic residue or quadratic nonresidue modulo p. This is called the **Legendre symbol**, after the French mathematician Adrien Marie Legendre (1752–1833).

Examples. From previous examples, we see that $(1/13) = (4/13) = (12/13) = 1$, while $(2/5) = (15/23) = -1$. ◊

Theorem 5.7. *Let p be an odd prime not dividing the integers a or b.*

1. $\left(\dfrac{a^2}{p}\right) = 1.$
2. $\left(\dfrac{ab}{p}\right) = \left(\dfrac{a}{p}\right)\left(\dfrac{b}{p}\right).$
3. $\left(\dfrac{a}{p}\right) \equiv a^{(p-1)/2} \pmod{p}.$
4. *If* $a \equiv b \pmod{p}$, *then* $\left(\dfrac{a}{p}\right) = \left(\dfrac{b}{p}\right).$
5. $\left(\dfrac{-1}{p}\right) = 1$ *or* -1 *according as* $p \equiv 1$ *or* $3 \pmod{4}$.

Proof. Parts (1) and (4) follow directly from the definition of quadratic residue. Part (3) is Euler's criterion, and part (5) is its corollary. Finally, part (2) follows from part (3) since

$$\left(\frac{a}{p}\right)\left(\frac{b}{p}\right) \equiv a^{(p-1)/2}b^{(p-1)/2} = (ab)^{(p-1)/2} \equiv \left(\frac{ab}{p}\right) \pmod{p}.$$

□

Problems for Section 5.3.

A

In the first four problems, list all the quadratic residues $m \pmod{p}$, $0 < m < p$.

1. $p = 7$ 2. $p = 13$ 3. $p = 17$ 4. $p = 19$

In the next 14 problems, evaluate the given Legendre symbol.

5. $(3/5)$
6. $(3/13)$
7. $(101/3)$
8. $(99/5)$
9. $(-10/13)$
10. $(-2/13)$
11. $(64/43)$
12. $(121/47)$
13. $(-1/53)$
14. $(-1/71)$
15. $(72/73)$
16. $(96/97)$
17. $(80/41)$
18. $(234/47)$

In the next six problems, find the least complete solution to the congruence. Start by making a table of the least residues of $1^2, 2^2, \ldots, 8^2 \pmod{17}$.

19. $2x^2 + 5x - 7 \equiv 0 \pmod{17}$
20. $4x^2 - 7x + 2 \equiv 0 \pmod{17}$
21. $3x^2 + 4x + 5 \equiv 0 \pmod{17}$
22. $3x^2 + 10x - 7 \equiv 0 \pmod{17}$
23. $x^2 + 9x + 3 \equiv 0 \pmod{17}$
24. $5x^2 + 11x - 2 \equiv 0 \pmod{17}$

Tell how many elements a complete solution to the congruence has in each of the next six problems.

25. $x^2 + 5x + 7 \equiv 0 \pmod{19}$
26. $x^2 + 7x + 2 \equiv 0 \pmod{41}$
27. $x^2 + 8x - 2 \equiv 0 \pmod{73}$
28. $x^2 + 3x - 5 \equiv 0 \pmod{29}$
29. $x^2 + 9x - 1 \equiv 0 \pmod{43}$
30. $x^2 + 9x + 1 \equiv 0 \pmod{41}$

In the next six problems, find the least complete solution.

31. $4x^2 + 6x + 1 \equiv 0 \pmod{19}$
32. $4x^2 + 6x + 1 \equiv 0 \pmod{11}$
33. $4x^2 + 6x + 1 \equiv 0 \pmod{13}$
34. $x^2 + 16x + 32 \equiv 0 \pmod{127}$
35. $5x^2 + 8x + 1 \equiv 0 \pmod{129}$
36. $3x^2 + 3x - 5 \equiv 0 \pmod{53}$

B

37. Finish the proof of the first theorem of this section by showing that if p is an odd prime not dividing a, if $y^2 \equiv b^2 - 4ac \pmod{p}$, and if $2ax \equiv -b + y \pmod{p}$, then $ax^2 + bx + c \equiv 0 \pmod{p}$. (**Hint:** Find u such that $2au \equiv 1 \pmod{p}$. Then $x \equiv u(-b + y) \pmod{p}$. Substitute this into the desired congruence.)
38. Show that if $m > 0$ and if a and b are integers such that $x^2 \equiv a \pmod{m}$ and $y^2 \equiv b \pmod{m}$, then $z^2 \equiv ab \pmod{m}$ is solvable.
39. Show that if p is an odd prime and $p \nmid ac$, then complete solutions to $ax^2 + bx + c \equiv 0 \pmod{p}$ and $cx^2 + bx + a \equiv 0 \pmod{p}$ have the same number of elements.
40. Suppose p is an odd prime not dividing ab. Show that $z^2 \equiv ab \pmod{p}$ is solvable exactly when both or neither of $x^2 \equiv a \pmod{p}$ and $y^2 \equiv b \pmod{p}$ are solvable.

In the next four problems, tell whether the statement is true and give a counterexample if it is false.

41. Let $m > 0$ and $(m, ab) = 1$. If neither $x^2 \equiv a \pmod{m}$ nor $y^2 \equiv b \pmod{m}$ is solvable, then $z^2 \equiv ab \pmod{m}$ is solvable.
42. If p is an odd prime, then complete solutions to $ax^2 + bx + c \equiv 0 \pmod{p}$ and $cx^2 + bx + a \equiv 0 \pmod{p}$ have the same number of elements.

43. If p is a prime congruent to 3 (mod 4) and $p \nmid a$, then exactly one of a and $-a$ is a quadratic residue (mod p).
44. If p is prime and $p \nmid a$, then the number of elements in a complete solution to $x^2 \equiv a \pmod{p^n}$ is 0 or 2.

C

45. Show that if $m > 0$ is the product of distinct odd primes and $(m, a) = 1$, then the number of elements in a complete solution to $x^2 \equiv a \pmod{m}$ is 0 or a power of 2.
46. Show that if p is an odd prime not dividing a, then a complete solution to $x^2 \equiv a \pmod{p^n}$ has the same number of elements for all positive integers n.
47. Show that if m is odd and $(m, a) = 1$, then the number of elements in a complete solution to $x^2 \equiv a \pmod{m}$ is 0 or a power of 2.

5.4 Quadratic Reciprocity

Identifying Quadratic Residues

The problem of telling whether an integer a is a quadratic residue modulo p or not occupied some of the greatest number theorists of the eighteenth century, including Euler, Lagrange, Legendre, and Gauss, who made the greatest contribution to the subject. To save words we will establish the following convention:

CONVENTION In this section, p will represent an odd prime not dividing the positive integer a. All congruences will be modulo p unless some other modulus is specified.

Note that we can assume a to be positive without loss of generality, since by part (2) of Theorem 5.7 $(-a/p) = (-1/p)(a/p)$, and $(-1/p)$ can be evaluated by part (5) of the same theorem.

About all we have to go on is Euler's criterion, which says

$$\left(\frac{a}{p}\right) \equiv a^{(p-1)/2},$$

and it would be nice to have some independent way to evaluate the expression on the right. In the proof of Euler's theorem (here particularized to the case of a prime modulus p), we evaluated a similar expression, showing that

$$a^{p-1} \equiv 1.$$

Recall the proof. As x runs through a reduced residue system modulo p so does ax, so

$$a^{p-1} \prod x = \prod (ax) \equiv \prod x,$$

where the products run over $x = 1, 2, \dots, p-1$. Dividing through by $\prod x$ yields Fermat's theorem.

In order to evaluate $a^{(p-1)/2}$ we need to employ a product with $(p-1)/2$ factors, say over $x = 1, 2, \dots, (p-1)/2$. Let us define the integer h (think of "half") to be $(p-1)/2$. The integers from 1 to $p-1$ then look like

$$1, 2, \dots, \frac{p-1}{2} = h, h+1 = \frac{p+1}{2}, \dots, p-1.$$

A proof like that for Fermat's theorem would have us look at the product

$$a^h \prod x = \prod (ax),$$

where x runs from 1 to h in the products.

The problem with this product is that for x between 1 and h, ax need not be congruent to some number in the same range. Certainly ax is congruent to one of the numbers

$$-h, -(h-1), \dots, -1, 0, 1, 2, \dots, h,$$

since these $2h+1=p$ integers comprise a complete residue system (mod p). In fact, we could leave out 0 as a possibility, since by our assumptions $p \nmid ax$.

Let us define x^* to be that unique number between $-h$ and h such that $ax \equiv x^*$, where x runs from 1 to h. Taking $p=7$ and $a=5$, for example, we compute the following table. Note that $h=(7-1)/2=3$.

x	1	2	3
ax	5	10	15
x^*	-2	3	1

Here is another example, with $p=13$ (so $h=6$) and $a=2$.

x	1	2	3	4	5	6
ax	2	4	6	8	10	12
x^*	2	4	6	-5	-3	-1

Notice that the values of x^* repeat those of x, except for some minus signs. This turns out always to be the case, which is the basis of the following theorem.

Theorem 5.8 (Gauss's lemma). *Let p be an odd prime not dividing the integer a. For $x=1,2,\dots,h=(p-1)/2$ let x^* be that integer congruent to ax (mod p) such that $-h \le x^* \le h$. Suppose exactly n values of x^* are negative. Then $(a/p)=(-1)^n$.*

Proof. First we show that the absolute values of the numbers x^* comprise the set $\{1,2,\dots,h\}$. Since these absolute values all fall into this set, and since there are exactly h of them, it suffices to show they are distinct, that is, if $x \neq y$, then $|x^*| \neq |y^*|$.

Suppose $|x^*|=|y^*|$, with x and y between 1 and h. If x^* and y^* have the same sign, then $x^*=y^*$, and so $ax \equiv ay$ and $x=y$ by the cancellation theorem. Otherwise $x^*=-y^*$, which means $ax \equiv -ay$. Thus $p \mid a(x+y)$. This is impossible because x and y are between 1 and h, so $2 \le x+y \le 2h = p-1$. Thus p cannot divide $x+y$, and by assumption $p \nmid a$.

The rest of the proof parallels that of Fermat's theorem. If we let x run from 1 to h in the products below, then by the definition of n

$$a^h \prod x = \prod(ax) \equiv \prod x^* \equiv (-1)^n \prod x.$$

Cancelling $\prod x$ gives $(-1)^n \equiv a^h$. But the latter is congruent to (a/p) by Euler's criterion. Since $p>2$ we must have $(-1)^n=(a/p)$. □

The Case $a = 2$

We will use Gauss's lemma to compute (a/p) for some specific values of a. It is important to remember that we don't really need to know exactly what n is, only whether it is odd or even. (This is called the **parity** of n.)

Let us see what happens when $a = 2$. We are to find numbers x^* congruent to the numbers

$$2, 4, 6, \ldots, 2h = p - 1$$

in the set $\{-h, -h+1, \ldots, h-1, h\}$, and count how many of them are negative. (One example of how this works is given for $p = 13$ earlier in this section.) In general, n is the number of integers x such that $p/2 < 2x < p$, as the following picture shows.

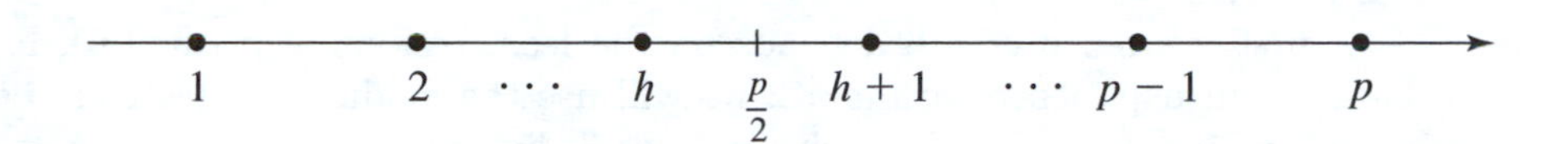

Thus we must count the number of integers x such that $p/4 < x < p/2$.

Of course, $p/4$ is not even an integer (nor is $p/2$), and so the smallest such x depends on what p is congruent to modulo 4. Let us suppose $p = 4q + r$, $0 \le r < 4$. Since p is odd we must have $r = 1$ or 3. Then we must count the integers x such that

$$\frac{p}{4} = q + \frac{r}{4} < x < 2q + \frac{r}{2} = \frac{p}{2}.$$

The first counted is clearly $q + 1$, and the last one counted is $2q$ or $2q + 1$, depending on whether $r = 1$ or 3. Thus $n = q$ or $q + 1$, according as $r = 1$ or 3.

The question is whether n is even or odd, which we see depends on both the parity of q and whether r is 1 or 3. Let us check cases.

case	$p = 4q + r$	n
$r = 1, q = 2s$	$8s + 1$	q, even
$r = 1, q = 2s + 1$	$8s + 5$	q, odd
$r = 3, q = 2s$	$8s + 3$	$q + 1$, odd
$r = 3, q = 2s + 1$	$8s + 7$	$q + 1$, even

Theorem 5.9. *Let p be an odd prime. Then $(2/p)$ is 1 if $p \equiv 1$ or 7 (mod 8), and $(2/p)$ is -1 if $p \equiv 3$ or 5 (mod 8).*

Example. Compute $(65/47)$.

Using various parts of Theorem 5.7 and the result just proved we have

$$\left(\frac{65}{47}\right) = \left(\frac{18}{47}\right) = \left(\frac{2}{47}\right)\left(\frac{9}{47}\right) = \left(\frac{2}{47}\right) = 1,$$

where the last equality holds since $47 \equiv 7 \pmod{8}$. ◊

The Case $a = 3$

A similar argument to the one previously can be made for $a = 3$. Since we are assuming $p \nmid a$, then $p > 3$. The multiples of $1, 2, \ldots, h$ are

$$3, 6, 9, \ldots, 3h = \frac{3(p-1)}{2},$$

all of which lie between 0 and $3p/2$. The following table shows what happens for $p = 17$ (so $h = 8$).

x	1	2	3	4	5	6	7	8
$3x$	3	6	9	12	15	18	21	24
x^*	3	6	-8	-5	-2	1	4	7

The negative values of x^* correspond to $p/2 < 3x < p$, which is equivalent to $p/6 < x < p/3$.

The first integer after $p/6$ depends on the least residue of $p \pmod 6$, but in light of our experience with $a = 2$ we will use the modulus 12 instead. Let $p = 12s + r$, $0 \le r < 12$. Since neither 2 nor 3 divides p, we must have $r = 1$, 5, 7, or 11.

Substituting $p = 12s + r$ into our inequality gives

$$2s + \frac{r}{6} < x < 4s + \frac{r}{3}.$$

Since shifting either endpoint of an interval by an even integer does not change whether the number of integers in the interval is even or odd, it suffices to count the number of integers y satisfying

$$\frac{r}{6} < y < \frac{r}{3}.$$

The number of such y will be even or odd the same as n. We resort to cases:

case	interval	possible y	parity of n
$r = 1$	$\frac{1}{6} < y < \frac{1}{3}$	none	even
$r = 5$	$\frac{5}{6} < y < 1\frac{2}{3}$	1	odd
$r = 7$	$1\frac{1}{6} < y < 2\frac{1}{3}$	2	odd
$r = 11$	$1\frac{5}{6} < y < 3\frac{2}{3}$	2, 3	even

Theorem 5.10. *Let $p > 3$ be prime. Then $(3/p) = 1$ if $p \equiv 1$ or 11 $\pmod{12}$ and $(3/p) = -1$ if $p \equiv 5$ or 7 $\pmod{12}$.*

Quadratic Reciprocity

The value of $(2/p)$ depends on what p is modulo 8. The value of $(3/p)$ depends on what p is modulo 12. Furthermore, some symmetry seems present. For example, $(2/p)$ has the same value whether $p \equiv 1 \pmod 8$ or $p \equiv -1 \equiv 7 \pmod 8$, and the same value whether $p \equiv 3 \pmod 8$ or $p \equiv -3 \equiv 5 \pmod 8$. The situation is similar for $a = 3$. Thus we might conjecture the following theorem.

Lemma 5.11. *Let $a > 0$, and let p and q be odd primes not dividing a. Then $(a/p) = (a/q)$ if $p \equiv q \pmod{4a}$ or if $p \equiv -q \pmod{4a}$.*

The proof of this lemma, which proceeds along lines similar to the cases $a = 2$ and $a = 3$ above, will be presented in Section 5.5. For the time being, we will assume it is true, and investigate its consequences. Suppose p and q are distinct odd primes, so that each is congruent to either 1 or 3 (mod 4). If $p \equiv q \pmod 4$, then, assuming $p < q$, we have $q - p = 4a$ for some positive integer a, and clearly neither p nor q divides a. Then $p \equiv q \pmod{4a}$, and so $(a/p) = (a/q)$ by Lemma 5.11. Since $q - p = 4a$, the integer a has the interesting property that

$$q \equiv 4a \pmod p \quad \text{and} \quad p \equiv -4a \pmod q.$$

Thus

$$\begin{aligned}\left(\frac{p}{q}\right) &= \left(\frac{-4a}{q}\right) = \left(\frac{-1}{q}\right)\left(\frac{4}{q}\right)\left(\frac{a}{q}\right) = \left(\frac{-1}{q}\right)\left(\frac{4}{p}\right)\left(\frac{a}{p}\right) \\ &= \left(\frac{-1}{q}\right)\left(\frac{4a}{p}\right) = \left(\frac{-1}{q}\right)\left(\frac{q}{p}\right),\end{aligned}$$

where this computation is justified by various parts of Theorem 5.7 (including $(4/q) = (4/p) = 1$ by part (1)) along with the equation $(a/p) = (a/q)$.

Recall that we know that $(-1/q)$ is 1 or -1 according as q (and so p also) is congruent to 1 or 3 (mod 4). We see that (p/q) and (q/p) are equal if both p and q are congruent to 1 (mod 4), but $(p/q) = -(q/p)$ if both are congruent to 3 (mod 4).

If $p \not\equiv q \pmod 4$, the situation is even simpler. Then $p + q = 4a$ for some positive integer a. Since $p \equiv -q \pmod{4a}$, Lemma 5.11 says that $(a/p) = (a/q)$. Now we have

$$p \equiv 4a \pmod q \quad \text{and} \quad q \equiv 4a \pmod p,$$

and so

$$\left(\frac{p}{q}\right) = \left(\frac{4a}{q}\right) = \left(\frac{4}{q}\right)\left(\frac{a}{q}\right) = \left(\frac{4}{p}\right)\left(\frac{a}{p}\right) = \left(\frac{4a}{p}\right) = \left(\frac{q}{p}\right).$$

We have proved the following celebrated theorem.

Theorem 5.12 (Gauss's law of quadratic reciprocity). *Suppose p and q are distinct odd primes. Then $(p/q) = (q/p)$ unless p and q are both congruent to 3 modulo 4, in which case $(p/q) = -(q/p)$.*

Examples. Evaluate (13/43), (19/59), and (37/67).

Since $13 \equiv 1 \pmod 4$ we have

$$\left(\frac{13}{43}\right) = \left(\frac{43}{13}\right) = \left(\frac{3 \cdot 13 + 4}{13}\right) = \left(\frac{4}{13}\right) = 1,$$

where we used, besides reciprocity, parts (4) and (1) of Theorem 5.7.

Likewise since both 19 and 59 are congruent to 3 (mod 4) we have

$$\left(\frac{19}{59}\right) = -\left(\frac{59}{19}\right) = -\left(\frac{2}{19}\right) = 1,$$

since $(2/19) = -1$ by Theorem 5.9.

Finally since $37 \equiv 1 \pmod 4$ we have

$$\begin{aligned}\left(\frac{37}{67}\right) &= \left(\frac{67}{37}\right) = \left(\frac{30}{37}\right) = \left(\frac{2}{37}\right)\left(\frac{3}{37}\right)\left(\frac{5}{37}\right) = (-1)\left(\frac{37}{3}\right)\left(\frac{37}{5}\right) \\ &= -\left(\frac{1}{3}\right)\left(\frac{2}{5}\right) = -(1)(-1) = 1,\end{aligned}$$

where Theorem 5.9 was used twice. Alternatively, we could note that $67 \equiv -7$ mod 37, so

$$\left(\frac{67}{37}\right) = \left(\frac{-7}{37}\right) = \left(\frac{-1}{37}\right)\left(\frac{37}{7}\right) = (1)\left(\frac{2}{7}\right) = 1.$$

◇

Problems for Section 5.4.

In the first six problems, use Theorems 5.7 and 5.9 to evaluate the given Legendre symbol.

1. (32/101) 2. (50/71) 3. (228/113)
4. (115/97) 5. (66/67) 6. (65/67)

In the next six problems, use Theorems 5.7, 5.9, and 5.10 to evaluate the given Legendre symbol, if possible.

7. (3/83) 8. (−3/89) 9. (77/71)
10. (600/131) 11. (27/91) 12. (162/53)

In the next 10 problems, evaluate the given Legendre symbol, using the law of quadratic reciprocity if necessary.

13. (17/67) 14. (19/67) 15. (53/73) 16. (86/103)
17. (1000/107) 18. (−1066/83) 19. (17/37) 20. (23/43)
21. (44/71) 22. (71/101)

In the next eight problems, determine whether the congruence has a solution.

23. $x^2 \equiv 15 \pmod{41}$ 24. $x^2 \equiv 1000 \pmod{157}$
25. $x^2 \equiv 234 \pmod{157}$ 26. $x^2 \equiv 158 \pmod{159}$

27. $x^2 - 3x + 17 \equiv 0 \pmod{61}$
28. $x^2 + 5x + 7 \equiv 0 \pmod{97}$
29. $3x^2 + 4x + 5 \equiv 0 \pmod{51}$
30. $4x^2 + 5x + 6 \equiv 0 \pmod{87}$

B

In the next four problems, tell the number of elements in a complete solution to the given congruence.

31. $x^2 \equiv 17 \pmod{143}$
32. $x^2 \equiv 23 \pmod{119}$
33. $3x^2 + 7x - 2 \equiv 0 \pmod{97}$
34. $2x^2 + 7x - 5 \equiv 0 \pmod{1099}$

35. Show that $(2/p) = (-1)^{(p^2-1)/8}$ for any odd prime p.
36. Show that the law of quadratic reciprocity is equivalent to the statement: If p and q are distinct odd primes, then $(p/q)(q/p) = (-1)^{(p-1)(q-1)/4}$.
37. Use the law of quadratic reciprocity to show that if p is an odd prime other than 5, then $(5/p) = 1$ if and only if the last digit of p is 1 or 9.
38. Make a table of $x = 1, 2, \ldots, h$ and x^* for $p = 17$ and $a = 5$.
39. Repeat the last problems for $p = 19$ and $a = 5$.
40. Show that if p is an odd prime other than 5 and r is the least residue of $p \pmod{20}$, then r is 1, 3, 7, 9, 11, 13, 17, or 19.
41. Find the number of integers y satisfying $r/10 < y < r/5$ for each r in the previous problem.
42. Find the number of integers y satisfying $3r/10 < y < 2r/5$ for each r in the second previous problem.
43. Prove that if $n > 0$, then the Fermat number F_n is congruent to 5 (mod 12).
44. Use Theorem 5.10 and Euler's criterion to prove the "only if" part of Pepin's test (Theorem 4.14).

C

The next three problems evaluate $(5/p)$, where p is an odd prime other than 5, in a manner like that used for $(2/p)$ and $(3/p)$ in this section. Do not use the law of quadratic reciprocity in these problems. Some of the work is already done in problems 40–42.

45. Show that if h and x^* are defined as in this section, then the number of x^* that are negative for $x = 1, 2, \ldots, h$ is the number n of integers satisfying $p/10 < x < p/5$ or $3p/10 < x < 2p/5$.
46. Show that if r is the least residue of $p \pmod{20}$, then the number n of the previous problem is even or odd according as the number of integers y satisfying either $r/10 < y < r/5$ or $3r/10 < y < 2r/5$ is even or odd.
47. Show that $(5/p)$ is 1 if $p \equiv 1$, 9, 11, or 19 (mod 20), and is -1 if $p \equiv 3$, 7, 13, or 17 (mod 20).

48. Show that the last problem agrees with the law of quadratic reciprocity.
49. Show that if $4a \mid p - q$, where p and q are odd primes, then $(a/p) = (a/q)$ even if a is negative.
50. Show that if $ab \equiv 1 \pmod{p}$, p an odd prime, then $(a/p) = (b/p)$.

5.5 Flipping a Coin over the Telephone

The Proof of Lemma 5.11

Our first order of business is to prove the lemma of the last section from which we derived the law of quadratic reciprocity.

Lemma 5.11. *Let $a > 0$, and let p and q be odd primes not dividing a. Then $(a/p) = (a/q)$ if $p \equiv q \pmod{4a}$ or if $p \equiv -q \pmod{4a}$.*

Proof. As with our evaluations of $(2/p)$ and $(3/p)$, we will employ Gauss's lemma. Let $h = (p-1)/2$, and consider the integers

$$a, 2a, 3a, \ldots, ha.$$

These fall into the open intervals

$$\left(0, \frac{p}{2}\right), \left(\frac{p}{2}, \frac{2p}{2}\right), \left(\frac{2p}{2}, \frac{3p}{2}\right), \left(\frac{3p}{2}, \frac{4p}{2}\right), \ldots, \tag{5.12}$$

where, as is customary, we are denoting the set of real numbers X such that $A < X < B$ by (A, B). Since

$$ha = \frac{(p-1)a}{2} < \frac{pa}{2} < \frac{(p+1)a}{2} = (h+1)a,$$

the last interval we need consider is $((a-1)p/2, ap/2)$. A total of a intervals are involved, so that the number of intervals does not depend on p.

Notice that the endpoints of the intervals listed in (5.12) are either nonintegers or else multiples of p. Thus none of the integers $a, 2a, \ldots, ha$ falls on one of these endpoints, since $p \nmid a$ and $h < p$.

As in Section 5.4, we define x^* by $x^* \equiv ax \pmod{p}$, $-h \leq x^* \leq h$. By Gauss's lemma the value of (a/p) depends on whether the number of negative x^* is even or odd. The integer x^* will be negative when ax falls in half the intervals listed in (5.12), namely, the intervals

$$\left(\frac{p}{2}, \frac{2p}{2}\right), \left(\frac{3p}{2}, \frac{4p}{2}\right), \left(\frac{5p}{2}, \frac{6p}{2}\right), \ldots.$$

Thus in a typical interval we want to count the number of integers x such that

$$\frac{(2k-1)p}{2} < ax < \frac{2kp}{2},$$

or

$$\frac{(2k-1)p}{2a} < x < \frac{2kp}{2a}. \tag{5.13}$$

Now assume q is an odd prime such that $q \equiv p \pmod{4a}$. Then $q = p + 4at$ for some integer t. If we try to evaluate (a/q) by Gauss's lemma in the same way, a typical interval in which we would be counting integers would be defined by the inequalities

$$\frac{(2k-1)q}{2a} < y < \frac{2kq}{2a}. \tag{5.14}$$

Plugging $q = p + 4at$ into this leads to

$$\frac{(2k-1)p}{2a} + (2k-1)2t < y < \frac{2kp}{2a} + 4kt. \tag{5.15}$$

(We leave it to the reader to check the algebra.)

If we compare the endpoints of the intervals defined by the inequalities (5.13) and (5.15) we see that the left endpoints differ by even integers, as do the right endpoints. Thus the number of integers in the corresponding intervals differs by a multiple of 2; it is even in both cases or odd in both cases. By using the same argument for each value of k and applying Gauss's lemma we conclude that $(a/p) = (a/q)$.

Now we consider the case when $q \equiv -p \pmod{4a}$. Then $q = -p + 4at$ for some integer t. Plugging this into (5.14) produces

$$\frac{-(2k-1)p}{2a} + (2k-1)2t < y < \frac{-2kp}{2a} + 4kt.$$

Multiplying through by -1 produces a symmetric interval on the other side of 0 that contains the same number of integers:

$$\frac{(2k-1)p}{2a} - (2k-1)2t > y' > \frac{2kp}{2a} - 4kt.$$

In fact, the same number of integers are in the interval shifted $4kt$ units to the right:

$$\frac{(2k-1)p}{2a} + 2t > y'' > \frac{2kp}{2a},$$

which can be written

$$\frac{2kp}{2a} < y'' < \frac{(2k-1)p}{2a} + 2t. \tag{5.16}$$

We would like to show that the number of integers y'' satisfying these inequalities is even or odd the same as the number of x satisfying (5.13). But (5.16) and (5.13) define adjacent intervals, and the number of integers in their union is the number of z satisfying

$$\frac{(2k-1)p}{2a} < z < \frac{(2k-1)p}{2a} + 2t.$$

(Recall that the endpoints of our interval are never hit so we need not worry about x equaling the common endpoint of the two intervals.)

The last inequalities define an interval of length $2t$ with nonintegral endpoints. It must contain an even number of integers. Thus the number of integers satisfying (5.13) and (5.16) must be even in both cases or odd in both cases. Again, using this argument for all values of k and applying Gauss's lemma, we see that $(a/q) = (a/p)$. □

An Application of Quadratic Residues

We conclude this chapter with an application, invented by Manuel Blum, that is similar to the RSA method of Section 4.5. Two people in distant cities want to settle a dispute by flipping a coin, but would like to avoid traveling to a common site.

Let us call the participants Al and Betty. Of course, Al and Betty don't have to be in the same room to flip a coin. Al could flip the coin, and Betty call heads or tails over the phone. Then Al tells Betty if her call is correct or not.

The obvious problem with this scenerio is that Al could lie, telling Betty she has lost no matter how the coin lands. We will describe a predetermined sequence of actions (or "protocol") that enables two people in different places to, in effect, flip a coin in a way that is fair to both.

As with the RSA method, the fairness of the system depends on the complexity of certain algorithms, a subject introduced in Section 4.4. Again at the core is the computational difficulty of factoring a large number, say of 300 digits, compared to other computations that are comparatively easy, at least for a computer.

In this exposition, we will take an informal approach, and not count elementary operations, but just talk of "easy" computations, meaning those that can be done, even for large numbers, in a reasonable amount of time on a computer. We list some easy computations below.

Euclidean algorithm This is easy by Lamé's theorem (Theorem 4.13).

Solving a linear congruence Section 4.5 tells how to use the quotients from applying the Euclidean algorithm to a and b to solve the congruence $ax \equiv c \pmod{b}$ efficiently.

Modular exponentiation We can compute $a^d \pmod{n}$ efficiently by Theorem 4.12.

The Protocol

We will describe the protocol so that Betty makes what corresponds to the call of heads or tails. She starts the process by choosing two distinct large primes p and q, each congruent to 3 (mod 4), and setting $n = pq$. (At present, taking p and q to be of about 150 digits seems sufficient.) Betty now tells Al the value of n, but keeps p and q secret.

Now the ball is in Al's court. He chooses an integer x relatively prime to n such that $0 < x < n$ and computes the least residue a of x^2 modulo n. (Al could check that x and n are relativly prime using the Euclidean algorithm, but actually the probability of his choosing for x a multiple of p or q is so small that this possibility can be ignored.) Al now tells Betty the value of a.

Now comes the part where Betty actually calls heads or tails, in effect. It turns out that the congruence

$$t^2 \equiv a \pmod{n}, \quad 0 < t < n, \tag{5.17}$$

has exactly four solutions t. (We will delay giving the reasons for this and various other mathematical statements while we explain the protocol.) Two of these are x and $n-x$, while the other two are of the form y and $n-y$. Because she knows how to factor n, Betty can compute all four solutions: x, $n-x$, y, and $n-y$. (How will be explained later.) She picks one of these.

Who wins? If Betty picks x (Al's original value) or $n-x$, she wins; but if she picks y or $n-y$, she loses.

Now Al tells Betty if she has won or not. If Betty has picked y or $n-y$, he proves to her she has picked wrong by revealing the value of x, which is different from y and $n-y$.

Example. Suppose Betty picks the primes $p = 31$ and $q = 43$. (We have scaled down the numbers in this example to allow hand calculator computations.) She computes $n = 31 \cdot 43 = 1333$ and tells this number to Al.

Now Al picks a number at random, say $x = 567$, and computes a, the least residue of $x^2 \pmod{n}$. Check that $567^2 = 321{,}489 \equiv 236 \pmod{1333}$. Thus $a = 236$. At this point Al knows two solutions to (5.17), or

$$t^2 \equiv 236 \pmod{1333}, \quad 0 < t < 1333, \tag{5.18}$$

namely $x = 567$ and $n - x = 1333 - 567 = 766$. All Al tells Betty, however, is that $a = 236$.

Now Betty find all the solutions to (5.18), by a method that will be explained below. They are $t = 164, 567, 766$, and 1169. Notice that $1169 = 1333 - 164$. Betty calls heads or tails by choosing one of the four solutions. If she chooses 567 or 766, the solutions Al knows, she wins, but if she chooses 164 or 1169, she loses. ◇

Why Al Can't Cheat

Al started with x, and can easily compute $n-x$, but because he doesn't know the factorization of n he cannot compute the other two solutions to (5.17). In fact, knowing more than two solutions is equivalent to factoring n. For suppose Al knows two solutions u and v to (5.17) with $u \not\equiv v$ and $u \not\equiv -v \pmod{n}$. Then

$$u^2 \equiv a \equiv v^2 \pmod{n},$$

and so

$$n \mid v^2 - u^2 = (v-u)(v+u).$$

Now if $n \mid v-u$ or $n \mid v+u$, then $u \equiv \pm v \pmod{n}$, contrary to our assumption. Since $n = pq$ with p and q prime we must have that p divides one of $v-u$ and q divides the other.

Since n does not divide $v - u$, this means that $(v - u, n)$ equals p or q. Al computes $(v - u, n)$ using the easy Euclidean algorithm, and winds up with one of the prime divisors of n; dividing this prime into n gives the other. We see that finding more than two solutions to (5.17) is equivalent to factoring n. But factoring an arbitrary 300-digit number is beyond the capability of any computer by presently known methods.

Example. Recall the previous example, where

$$t^2 \equiv 236 \pmod{1333}, \quad 0 < t < 1133$$

has the solutions $t = 164, 567, 766, 1169$. These form pairs 164, 1169 and 567, 766, since $164 + 1169 = 567 + 766 = 1333$. Suppose that Al knows solutions from different pairs, say $u = 164$ and $v = 766$. Al uses the Euclidean algorithm to compute $(v - u, n) = (766 - 164, 1333) = (602, 1333)$ as follows.

$$\begin{aligned} 1333 &= 602 \cdot 2 + 129 \\ 602 &= 129 \cdot 4 + 86 \\ 129 &= 86 \cdot 1 + 43 \\ 86 &= 43 \cdot 2 + 0 \end{aligned}$$

The last nonzero remainder is 43, and sure enough $1333 = 43 \cdot 31$. Al has used his knowledge of solutions from different pairs to factor n. ◇

How Betty Gets Her Four Solutions

That (5.17) has exactly four solutions is easy to see by what we know about quadratic congruences. Since $x^2 \equiv a \pmod{n}$, the same congruence holds modulo both p and q. Thus a is a quadratic residue modulo both primes, and a complete solution to $t^2 \equiv a \pmod{p}$ has exactly two elements; likewise for the modulus q. By the Chinese remainder theorem each pair of solutions modulo p and modulo q leads to one solution modulo $pq = n$, so a complete solution modulo n has exactly four elements.

Of course, knowing there are four solutions does not tell Betty how to find them. But for her this problem is reduced to the problems of solving

$$t^2 \equiv a \pmod{p} \quad \text{and} \quad t^2 \equiv a \pmod{q}.$$

Then the four solutions modulo n could be found using the Chinese remainder theorem. True, applying the theorem involves solving linear congruences modulo p and q, but we have already noted that solving such congruences is an easy computation.

It turns out that because $p = 4k + 3$ for some integer k, solving

$$t^2 \equiv a \pmod{p}$$

is also easy. (You may have forgotten, but the condition that p and q be congruent to 3 (mod 4) was in the protocol.) In fact, $t = a^{k+1}$ is a solution. For

$$(a^{k+1})^2 = a \cdot a^{2k+1} \equiv a(x^2)^{2k+1} = ax^{4k+2} = ax^{p-1} \equiv a \pmod{p}$$

by Fermat's theorem.

Note that the modular exponentiation algorithm takes care of the computation of a^{k+1}. Similarly, if $q = 4j + 3$, then a^{j+1} is an explicit solution to $t^2 \equiv a \pmod{q}$.

Example. Let us show how Betty solves

$$t^2 \equiv 236 \pmod{1333}, \quad 0 < t < 1333,$$

by knowing that $1333 = 31 \cdot 43$. Note that both 31 and 43 are congruent to 3 (mod 4).

First she solves

$$t^2 \equiv 236 \equiv 19 \pmod{31}.$$

Note that $31 = 4 \cdot 7 + 3$, so $k = 7$. She must compute $a^{k+1} \equiv 19^8 \pmod{31}$. Check that $19^2 = 361 \equiv 20$, $19^4 \equiv 20^2 = 400 \equiv 28$, and $19^8 = 28^2 = 784 = 9 \pmod{31}$. Thus $t = 9$ is a solution; another is $t = -9$.

Likewise, Betty solves

$$t^2 \equiv 236 \equiv 21 \pmod{43}.$$

Since $43 = 4 \cdot 10 + 3$, $j = 10$. She must compute $a^{j+1} \equiv 21^{11} \pmod{43}$. Check that $21^2 = 441 \equiv 11$, $21^4 \equiv 11^2 = 121 \equiv 35$, and $21^8 \equiv 35^2 = 1225 \equiv 21 \pmod{43}$. Now $21^{11} = 21^{1+2+8} = 21^1 21^2 21^8 \equiv 21 \cdot 11 \cdot 21 = 4851 \equiv 35 \pmod{43}$. Thus $t = 35$ is a solution, another is $t = -35$.

Now Betty uses the Chinese remainder theorem to solve the simultaneous congruences

$$t \equiv \pm 9 \pmod{31} \quad \text{and} \quad t \equiv \pm 35 \pmod{43}.$$

This involves solving the linear congruences

$$43x_1 \equiv 1 \pmod{31} \quad \text{and} \quad 31x_2 \equiv 1 \pmod{43}.$$

We will skip the details but ask the reader to check that $x_1 = 13$ and $x_2 = -18$ are solutions. Thus we have

$$t = 43 \cdot 13(\pm 9) + 31(-18)(\pm 35) = -24{,}561, -14{,}499, 14{,}499, 24{,}561.$$

Taking least residues (mod 1333) gives the complete solution $t = 164$, 567, 766, 1169. ◇

Why Betty Can't Cheat

Betty might try to cheat Al by giving him a value of n that was prime. Then (5.17) would have only two solutions, and she would be sure of picking one that Al knew. If Al loses, he should demand that Betty tell him the factors p and q of n to make sure he wasn't flipping a coin with two heads.

Problems for Section 5.5.

A

In the first four problems, suppose Betty chooses p and q, then Al picks x, and later Betty picks x'. Who wins, Al or Betty?

1. $p = 43, q = 19, x = 145, x' = 102$
2. $p = 31, q = 23, x = 302, x' = 302$
3. $p = 47, q = 59, x = 1007, x' = 1766$
4. $p = 11, q = 67, x = 466, x' = 70$

In the next four problems, let p, q, and x be as in the protocol for flipping a coin over the telephone. Find a.

5. $p = 19, q = 31, x = 143$
6. $p = 19, q = 43, x = 501$
7. $p = 47, q = 23, x = 392$
8. $p = 23, q = 43, x = 717$

In the next four problems, a prime $p = 4k + 3$ is given. Find the least complete solution to $t^2 \equiv a \pmod{p}$ by the method of this section.

9. $p = 23, a = 6$
10. $p = 19, a = 11$
11. $p = 43, a = 31$
12. $p = 47, a = 42$

Given that $t_1^2 \equiv a \pmod{p}$ and $t_2^2 \equiv a \pmod{q}$, find the least complete solution to $t^2 \equiv a \pmod{pq}$.

13. $p = 19, q = 23, a = 163, t_1 = 12, t_2 = 18$
14. $p = 31, q = 19, a = 524, t_1 = 20, t_2 = 12$
15. $p = 23, q = 43, a = 547, t_1 = 8, t_2 = 17$
16. $p = 47, q = 11, a = 504, t_1 = 38, t_2 = 8$

B

In the next four problems, use the method of this section to find the least complete solution to $t^2 \equiv a \pmod{pq}$.

17. $p = 11, q = 7, a = 15$
18. $p = 17, q = 19, a = 92$
19. $p = 31, q = 11, a = 152$
20. $p = 19, q = 23, a = 418$
21. Show that in the proof of Lemma 5.11 the maximum value of k in (5.13) is $a/2$ if a is even and $(a - 1)/2$ if a is odd.
22. Consider the proof of Lemma 5.11 when $a = 7, p = 11$, and $q = 67$, so $p \equiv q \pmod{4a}$. Count the number of integers satisfying (5.13) for $k = 1$, 2, and 3. Do the same for (5.14), and confirm that corresponding intervals both contain even or odd numbers of integers.

23. Repeat the previous problem, but change q to 73. Then $p \equiv -q \pmod{4a}$, and in fact $p+q = 4at$ with $t = 3$. Count the integers satisfying (5.14) for $k = 1, 2$, and 3 and confirm that for each k the total number of integers satisfying (5.13) or (5.14) is $2t = 6$.
24. Suppose $p > 3$ is prime. Show that $(-3/p) = 1$ if and only if $p \equiv 1 \pmod{3}$.
25. Let p be an odd prime. When does $x^2 \equiv (2/p) \pmod{p}$ have no solutions?

C

26. Let p be an odd prime other than 7. When is $x^2 \equiv (7/p) \pmod{p}$ without solutions?

G. H. Hardy and Srinivasa Ramanujan

1877–1947 1887–1920

Hardy hated to have his picture taken and used towels to cover all the mirrors in hotel rooms to avoid seeing his own image. He called God his personal enemy, refusing to enter any religious building. Ramanujan was a vegetarian who always changed into pajamas before cooking his own meals. He agreed to travel from India to England only after his mother had a dream in which the goddess Namagiri approved the journey. The personal idiosyncrasies of Hardy and Ramanujan are no more remarkable, however, than the mathematical chain of events that brought them together.

G. H. Hardy was a Cambridge professor who defended his colleague Bertrand Russell's antiwar activities during World War I. Hardy's specialty was analytic number theory, an area in which he and his collaborator, J. E. Littlewood, made fundamental advances. Hardy was a witty conversationalist who loved tennis and cricket, and adopted baseball when visiting the United States. His small book *A Mathematician's Apology* lays out in graceful prose his life in the purest of pure mathematics.

One day in 1913 Hardy received a letter from India full of mathematical formulas. Some were standard results, but others were like nothing he had seen before. Hardy recognized the genius of the writer of the letter, Srinivasa Ramanujan, and immediately arranged for him to come to Cambridge.

Ramanujan was born in the Tanjore district of Madras, the son of an accountant to a cloth merchant. At 15, Ramanujan gained access to Carr's *Synopsis of Pure Mathematics*, a compendium of formulas of elementary mathematics. Ramanujan used this book to teach himself the subject, and soon he was discovering mathematical laws on his own. Lacking formal mathematical training, he proceed on a basis of intuition and experiment that is hard to reconstruct. According to Hardy, Ramanujan lacked the modern idea of a proof.

When Ramanujan got to England, Hardy found in him a genius who was ignorant of many fundamental areas of mathematics and had the problem of acquainting him with modern ideas without stifling his originality. They wrote a series of papers together.

The British climate did not agree with Ramanujan, and he developed tuberculosis. When Hardy visited him in a hospital to make conversation he mentioned that the number of his taxicab had been 1729, "a rather dull number." Ramanujan replied "No Hardy! No Hardy! It is a very interesting number. It is the smallest number expressible as the sum of two cubes in two different ways." Ramanujan returned to India for his health, but died soon after, at the age of 32.

G. H. HARDY

SRINIVASA RAMANUJAN

Chapter 6

The Number Theory of the Reals

Although there are many real numbers, rational and irrational, that are not integers, the integers may be important in studying these numbers. Connections between the integers and nonintegers will be investigated in this chapter.

6.1 Rational and Irrational Numbers

Countability

That not all real numbers are rational, that is, the quotient of two integers, was known to ancient Greek mathematicians. We looked at the usual proof that $\sqrt{2}$ is irrational in Chapter 0. With what we know now about prime factorization, that proof can be generalized considerably. It turns out that $\sqrt[k]{n}$ is irrational whenever n is not a perfect kth power.

Theorem 6.1. *Let n and k be positive integers. Then if $\sqrt[k]{n}$ is not an integer, it is irrational.*

Proof. Suppose $\sqrt[k]{n}$ is rational, say $\sqrt[k]{n} = a/b$, where a and b are positive integers. Then

$$b^k n = a^k.$$

Let us consider the factorizations of a, b, and n into prime powers. Suppose the prime p occurs in the factorizations of a, b, and n to the powers r, s, and t, respectively. Since p appears the same number of times in the factorizations of both sides of our equation, we have

$$(p^s)^k p^t = (p^r)^k, \quad \text{so} \quad sk + t = rk.$$

We see that t, the power of p occuring in n, must be a multiple of k. Since this holds for each prime dividing n, we can conclude that n is a perfect kth power, and so $\sqrt[k]{n}$ is an integer. $\square$

This theorem lets us exhibit as many irrational numbers as we wish, for example, $\sqrt{3}, \sqrt{5}, \sqrt{6}, \sqrt[3]{3}$, and $\sqrt[13]{7}$. As a matter of fact, there are more irrational numbers than rational, but some definitions need to be given to make this statement meaningful. There are infinitely many rational number and infinitely many irrational, so what is needed is a way of comparing infinite sets.

DEFINITION. countable, uncountable

We say a set S is **countable** if it can be put into a one-to-one correspondence with the set of positive integers. This simply means there exists a rule that matches each element in S with a unique positive integer, pairing each positive integer with exactly one element of S. (Other common terminology for this situation is that there is a "one-to-one function from S onto the positive integers" or a "bijection from S to the positive integers.") If a set is infinite but not countable, then we say that it is **uncountable**.

Examples. First, we show that the set of even positive integers is countable. The required one-to-one correspondence is simply that matching 2 with 1, 4 with 2, 6 with 3, and, in general, the even integer $2n$ with n. The following table show how the matching starts.

even positive integer	2	4	6	8	10	12	14
positive integer	1	2	3	4	5	6	7

There is something somewhat paradoxical about this matching, since it indicates that in some sense there are "the same number" of even positive integers as positive integers, in spite of the fact that the first set is a proper subset of the second. After all, two finite sets have the same number of elements if and only if they can be matched up in a one-to-one fashion.

As another example, we will show that the set of all integers is countable. This time we will match 0 with 1, 1 with 2, -1 with 3, 2 with 4, -2 with 5, etc. The following table shows how the matching continues.

integer	0	1	-1	2	-2	3	-3	4	-4	5
positive integer	1	2	3	4	5	6	7	8	9	10

In this example, it is not so easy to give a simple algebraic rule for the matching. An algebraic formula is not necessary, however, so long as it is clear the set is matched with the positive integers in a one-to-one fashion.

Uncountable sets also exist, but we will not give an example of one until later in this section. ◇

Clearly any countable set is infinite, since it can be matched with the infinite set of positive integers. In some sense, the smallest infinite sets are those that are countable. The following theorem makes this statement more precise.

Theorem 6.2. *Let S be any infinite set. Then S contains a countable subset.*

Proof. Choose an element x_1 from S. Let S_1 be formed by removing x_1 from S. Now choose x_2 from S_1. Let S_2 be the set formed by removing x_2 from S_1.

Choose x_3 from S_2. In general, we choose x_{i+1} from S_i and then define S_{i+1} to be S_i with x_{i+1} removed.

Notice that if we continue in this way, each set S_i is infinite because it differs from S by a finite number of elements. Thus the process will never end and we can define an infinite sequence of distinct elements $x_1, x_2, \ldots$ of S this way. Let T be the set of all these elements $x_1, x_2, \ldots$. Then matching x_i with the positive integer i shows that T is countable. $\square$

Note that the proof just given does not show that S itself is countable, since there may be elements of S that never get put into T. Of course, we can put any *particular* element of S into T if we want to, but this does not mean that we can exhaust S this way. If S is uncountable then it cannot equal T, which we just showed to be countable. In truth, we have not exhibited a single uncountable set yet. Actually, the set of real numbers is uncountable, as we will see in the next theorem. First, we will introduce a notation for an idea that was used back in Section 1.2 in the proof of the division algorithm, namely, writing a real number as the sum of an integer and a number between 0 and 1.

DEFINITION. integer part, fractional part

Let X be any real number. By the **integer part** of X we mean the greatest integer n such that $n \leq X$. We denote this integer n by $\lfloor X \rfloor$. For example, $\lfloor 3.6 \rfloor = 3$, $\lfloor \pi \rfloor = 3$, $\lfloor 17 \rfloor = 17$, and $\lfloor -3.68 \rfloor = -4$. The **fractional part** of X is defined to be $X - \lfloor X \rfloor$. Thus the fractional part of 3.6 is .6, the fractional part of -3.68 is $-3.68 - (-4) = .32$, the fractional part of 17 is 0, and the fractional part of $\sqrt{2}$ is the (irrational) number .41421356 .... It is easy to see that the fractional part of every real number is greater than or equal to 0 and strictly less than 1.

An Uncountable Set

It may be hard to see how one would prove a set was *not* countable, since this would mean showing that no one-to-one correspondence between the set and the positive integers existed. What is called for is a proof by contradiction. The following famous "diagonal argument" is by the German mathematician Georg Cantor, who developed much of the theory of infinite sets.

Theorem 6.3. The set of real numbers is uncountable.

Proof. Denote by R the set of real numbers. Let us suppose R is countable. Then there exists a one-to-one matching between R and the positive integers. Now some element of R is matched with 1; let us call it X_1. In general let X_i denote the element of R matched with the positive integer i. Then by the definition of a one-to-one correspondence the numbers

$$X_1, X_2, X_3, \ldots$$

comprise all real numbers.

Each of these real numbers may be written as the integer $\lfloor X_i \rfloor$ plus its fractional part, where the latter is greater than or equal to 0 and less than 1. We will represent this fractional part as a decimal. Let us suppose

$$X_1 = \lfloor X_1 \rfloor + 0.a_{11}a_{12}a_{13}\ldots,$$
$$X_2 = \lfloor X_2 \rfloor + 0.a_{21}a_{22}a_{23}\ldots,$$

and, in general,

$$X_i = \lfloor X_i \rfloor + 0.a_{i1}a_{i2}a_{i3}\ldots,$$

where each a_{ij} is a digit from 0 to 9.

Now we will derive a contradiction by showing that there is an element of R that is not one of the numbers X_i. This is the number Y with the decimal representation

$$Y = 0.b_1b_2b_3\ldots,$$

where the digits b_i are defined as follows:

$$b_i = \begin{cases} 7 & \text{if } a_{ii} = 0, 1, 2, 3, \text{ or } 4 \\ 4 & \text{if } a_{ii} = 5, 6, 7, 8, \text{ or } 9. \end{cases}$$

The number Y is defined so as to differ from a_1 in its first digit to the right of the decimal point, from a_2 in its second digit to the right of the decimal point, etc.

Just to illustrate the definition of Y, let us suppose the first few numbers $X_1, X_2, \ldots$ are as follows.

$$\begin{array}{rrl} X_1 = & 17 & +\,.\textcircled{0}\,7\,3\,2\,9\,0\,0\ldots \\ X_2 = & 3 & +\,.5\,\textcircled{6}\,3\,7\,4\,5\,5\ldots \\ X_3 = & -6 & +\,.1\,9\,\textcircled{7}\,4\,4\,6\,2\ldots \\ X_4 = & 0 & +\,.8\,0\,0\,\textcircled{1}\,7\,8\,3\ldots \\ X_5 = & 594 & +\,.2\,4\,7\,6\,\textcircled{7}\,9\,6\ldots \end{array}$$

The digits important for the definition of the digits b_i are circled. Then $a_{11} = 0$, so $b_1 = 7$. Likewise $a_{22} = 6$, so $b_2 = 4$. In this example we see that Y starts out

$$Y = .74474\ldots.$$

The contradiction is now that Y is in R but Y is not matched with any positive integer, contrary to our assumption. For if Y were matched with the positive integer i, then we would have $Y = X_i$. But Y differs from X_i in its ith digit, and so cannot equal X_i. This contradiction shows that R is uncountable. □

Infinite sets have many properties that may at first seem strange. Putting two of them together, for example, does not necessarily yield more elements than are in just one of the sets.

Theorem 6.4. *Let S and T be countable sets with no elements in common. Then the union of S and T (that is, the set of all elements in either set) is also countable.*

Proof. Since S is countable, its elements can be matched with the positive integers in a one-to-one fashion. Let X_1 be the element of S matched with 1, X_2 be the element matched with 2, etc. Then

$$S = \{X_1, X_2, \dots\}.$$

Likewise let T consist of the distinct elements $Y_1, Y_2, \dots$. We match the elements of the union of S and T with the positive integers as indicated in the following table.

element	X_1	Y_1	X_2	Y_2	X_3	Y_3	X_4	Y_4	...
positive element	1	2	3	4	5	6	7	8	...

This matching shows that the union of the sets is countable. □

Theorem 6.5. *The set of rational numbers is countable.*

Proof. We need to match the rational numbers with the positive integers. The following array, which lists first those rationals with denominator 1, then those with denominator 2, etc., is the key to our method.

$$\begin{array}{ccccccccc}
\frac{0}{1}, & \frac{1}{1}, & \frac{-1}{1}, & \frac{2}{1}, & \frac{-2}{1}, & \frac{3}{1}, & \frac{-3}{1}, & \frac{4}{1}, & \cdots \\
\frac{0}{2}, & \frac{1}{2}, & \frac{-1}{2}, & \frac{2}{2}, & \frac{-2}{2}, & \frac{3}{2}, & \frac{-3}{2}, & \frac{4}{2}, & \cdots \\
\frac{0}{3}, & \frac{1}{3}, & \frac{-1}{3}, & \frac{2}{3}, & \frac{-2}{3}, & \frac{3}{3}, & \frac{-3}{3}, & \frac{4}{3}, & \cdots \\
\frac{0}{4}, & \frac{1}{4}, & \frac{-1}{4}, & \frac{2}{4}, & \frac{-2}{4}, & \frac{3}{4}, & \frac{-3}{4}, & \frac{4}{4}, & \cdots
\end{array}$$

Of course, there are repeats in this table, but these will be eliminated when we define our matching with the positive integers. Our matching will follow the pattern shown below.

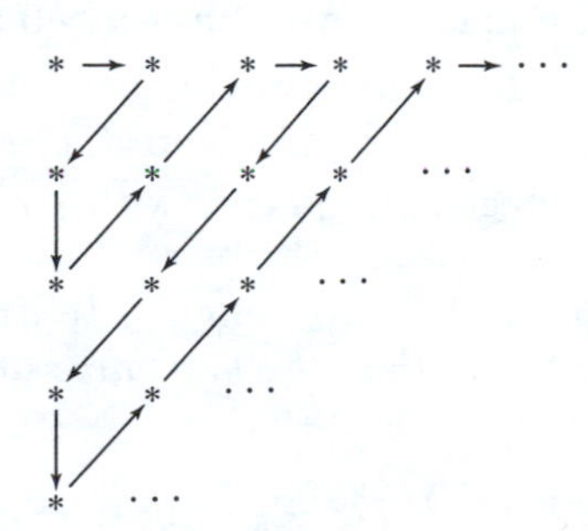

We match the rational numbers with the positive integers according to this pattern, skipping any rationals we have already used. The correspondence starts as follows.

rational	$\frac{0}{1}$	$\frac{1}{1}$	$\frac{1}{2}$	$\frac{-1}{1}$	$\frac{2}{1}$	$\frac{-1}{2}$	$\frac{1}{3}$	$\frac{1}{4}$	$\frac{-1}{3}$	$\frac{-2}{1}$
positive integer	1	2	3	4	5	6	7	8	9	10

This one-to-one correspondence demonstrates that the rational numbers are countable. □

Corollary 6.6. *The set of irrational numbers is uncountable.*

Proof. We have just seen that the rational numbers form a countable set. If the irrationals were also countable, then the real numbers, being the disjoint union of two countable sets, would also be countable by our theorem about such unions. But we know the real numbers are uncountable. □

Problems for Section 6.1.

A

In the first eight problems, compute the integer part and fractional part of the given number.

1. 64.716
2. 13/7
3. -4.333
4. $-7/11$
5. $(\sqrt{13}-1)^2+2(\sqrt{13}-1)$
6. $-3+.012$
7. $4-.01$
8. $10^{1/3}$

In the next eight problems, tell whether the indicated rule gives a one-to-one correspondence between the set S and the positive integers. If it does not, tell what is wrong.

9. S is the nonnegative integers; n matches with $n+1$.
10. S is the positive integers; n matches with n^2.
11. S is the positive integers congruent to 2 (mod 3); $3n+2$ matches with n.
12. S is the positive integers; n matches with the least residue of n^2 modulo $n+3$.
13. S is the integers; n matches with $2n$ if $n>0$ and $1-2n$ otherwise.
14. S is the integers; n matches with $2|n|+n+1$.
15. S is the odd positive integers; if n is in S, then n matches with $(n+1)/2$.
16. S is the positive integers; n matches with $\lfloor (n^2+1)^{1/2} \rfloor$.

In the next two problems, let X_1, X_2, and X_3 be as in the proof that the set of real numbers is uncountable. Find the first three digits of Y, where Y is as in that proof.

17. $X_1 = .12357\ldots, X_2 = .61728, X_3 = .98765\ldots$

18. $X_1 = 1/3, X_2 = \sqrt{2}, X_3 = 2^{-10}$

The next two problems refer to the proof that the rational numbers are countable.

19. What rational numbers match with the positive integers 11 and 12?
20. What rational numbers match with the positive integers 13 and 14?

B

21. Show that the theorem about the union of two countable sets is valid even if the sets may have elements in common.
22. Some numbers may be represented as a decimal in more than one way. For example, $.4000\ldots = .3999\ldots$. In the proof that the reals are uncountable, any decimal representation may be used for the fractional parts of the numbers X_i. Show that this ambiguity does not invalidate the proof given.

In the next four problems, show the given set S is countable by defining a one-to-one correspondence between it and the positive integers.

23. S is all positive integers congruent to 3 (mod 5)
24. $S = \{1, 4, 9, 16, 25, \ldots\}$
25. $S = \{1, 2, 4, 8, 16, \ldots\}$
26. S is all integers congruent to 2 (mod 4)

27. Show that if X is any real number, then $X - 1 < \lfloor X \rfloor \leq X$.

True-False. In the next eight problems, tell which statements are true and give counterexamples for those that are false. Assume that X, Y, and Z are real numbers and n is an integer.

28. $\lfloor X + Y \rfloor = \lfloor X \rfloor + \lfloor Y \rfloor$
29. $\lfloor X + n \rfloor = \lfloor X \rfloor + n$
30. $\lfloor -X \rfloor = -\lfloor X \rfloor$
31. $\lfloor -X \rfloor = -\lfloor X \rfloor - 1$
32. $\lfloor X \rfloor \leq X < \lfloor X \rfloor + 1$
33. $\lfloor X \rfloor + \lfloor Y \rfloor \leq \lfloor X + Y \rfloor \leq \lfloor X \rfloor + \lfloor Y \rfloor + 1$
34. $\lfloor X + Y + Z \rfloor \leq \lfloor X \rfloor + \lfloor Y \rfloor + \lfloor Z \rfloor + 1$
35. If $\lfloor X \rfloor = \lfloor Y \rfloor$, then $|X - Y| < 1$.

36. Show that if A and B are real numbers such that $A < B$, then there exists a rational number between A and B.
37. Do the previous problem with "rational" replaced by "irrational".

C

38. Suppose $S_1, S_2, \ldots, S_n$ are countable sets. Show that their union is countable.
39. Let S_i be a countable set for each positive integer i. Show that the union of all the sets S_i is countable.
40. Let S be the set of all finite subsets of the positive integers. Show that S is countable. (**Hint:** For each n there are finitely many subsets of $\{1, 2, \ldots, n\}$. Adapt the proof that the rationals are countable.)

41. Let S be the set of all subsets of the positive integers. Show that S is uncountable. (**Hint:** Associate each element T of S with a sequence of 0s and 1s, $a_1, a_2, \ldots$, where $a_i = 1$ if and only if i is in T. Adapt the proof that the reals are uncountable.)
42. Let $e = 1 + 1/1! + 1/2! + 1/3! + \cdots$. Show that if k is a positive integer, then $k!e$ is not an integer. (**Hint:** Show that $(k+1)^{-1} + (k+1)^{-1}(k+2)^{-1} + (k+1)^{-1}(k+2)^{-1}(k+3)^{-1} + \cdots < 1$ by comparing the expression on the left to a geometric series.)
43. Show that e is irrational. (See the previous problem.)
44. A real number is **algebraic** if it satisfies an equation $P(x) = 0$, where P is a nonconstant polynomial with integer coefficients. Show that the set of real numbers satisfying such an equation with the degree of P not more than n is countable for each positive integer n. (**Hint:** Problem 39.)
45. Show that the set of algebraic numbers is countable, and thus there exist **transcendental** numbers, i.e., real numbers that are not algebraic.

6.2 Finite Continued Fractions

The Euclidean Algorithm Revisited

In Section 1.3 we learned how to apply the Euclidean algorithm to the integers a and b to find (a, b), then solve the equations backward to find x and y such that $ax + by = (a, b)$. The process of solving the equations backward is cumbersome, invites mistakes, and is hard to program, so we will consider another method here.

Recall that the Euclidean algorithm equations have the following form.

$$\begin{aligned} b &= aq_1 + r_1 \\ a &= r_1 q_2 + r_2 \\ r_1 &= r_2 q_3 + r_3 \\ &\cdots \\ r_{i-2} &= r_{i-1} q_i + r_i \\ &\cdots \end{aligned}$$

Our idea will be to use the first equation to write r_1 as a linear combination of a and b, then substitute this into the second equation to write r_2 as a linear combination of a and b, and so on, working from the top down instead of from the bottom up. For each i we seek integers x_i and y_i such that

$$r_i = ax_i + by_i.$$

For example, from the first equation we have $r_1 = a(-q_1) + b$, so we take

$$x_1 = -q_1, \quad y_1 = 1. \tag{6.1}$$

Likewise from the second equation, we have

$$\begin{aligned} r_2 &= a - r_1 q_2 = a - (-aq_1 + b)q_2 \\ &= a(1 + q_1 q_2) + b(-q_2), \end{aligned}$$

so we take

$$x_2 = 1 + q_1 q_2, \quad y_2 = -q_2. \tag{6.2}$$

Assume we have found x_k and y_k for $k = 1, 2, \ldots, i - 1$. Then from the Euclidean algorithm equation with remainder r_i, we need

$$\begin{aligned} r_i &= r_{i-2} - r_{i-1} q_i \\ &= ax_{i-2} + by_{i-2} - (ax_{i-1} + by_{i-1})q_i \\ &= a(x_{i-2} - q_i x_{i-1}) + b(y_{i-2} - q_i y_{i-1}). \end{aligned}$$

Thus we need to have

$$x_i = x_{i-2} - q_i x_{i-1}, \quad y_i = y_{i-2} - q_i y_{i-1}. \tag{6.3}$$

We can use (6.1) and (6.2) to define x_i and y_i for $i = 1$ and 2, and then calculate the values for larger i from (6.3). Since the formulas of (6.1) and (6.2) are hard to remember, we will define "phony" values of x_i and y_i for $i = 0$ and -1, consistent with (6.3), which will be simpler and from which the $i = 1$ and 2 values can be computed.

In particular, we want to define x_0 and y_0 so that

$$x_2 = 1 + q_1 q_2 = x_0 - q_2 x_1 = x_0 + q_2 q_1,$$

and

$$y_2 = -q_2 = y_0 - q_2 y_1 = y_0 - q_2,$$

from which we see that we can take $x_0 = 1$ and $y_0 = 0$.

In the same way, we want to define x_{-1} and y_{-1} so that

$$x_1 = -q_1 = x_{-1} - q_1 x_0 = x_{-1} - q_1,$$

and

$$y_1 = 1 = y_{-1} - q_1 y_0 = y_{-1},$$

so we will take $x_{-1} = 0$ and $y_{-1} = 1$.

Theorem 6.7. *Let the Euclidean algorithm be applied to integers a and b, $a > 0$, producing quotients q_i and remainders r_i. Define integers x_i and y_i by*

$$\begin{aligned} x_{-1} &= 0, \quad x_0 = 1, \qquad x_i = x_{i-2} - q_i x_{i-1} \text{ for } i \geq 1, \\ y_{-1} &= 1, \quad y_0 = 0, \qquad y_i = y_{i-2} - q_i y_{i-1} \text{ for } i \geq 1. \end{aligned}$$

Then $r_i = ax_i + by_i$ for $i > 0$. In particular, if r_n is the last nonzero remainder, then $(a, b) = ax_n + by_n$.

Example. Let us take $a = 158$ and $b = 188$. In Section 1.3, the Euclidean algorithm was applied to these numbers, producing quotients 1, 5, 3, 1, and 3, and remainders 30, 8, 6, 2, and 0. (The reader should review this calculation.) Thus $(158, 188) = r_4 = 2$. We will set up a table to compute the values of x_i and y_i.

i	q_i	$x_i = x_{i-2} - q_i x_{i-1}$	$y_i = y_{i-2} - q_i y_{i-1}$
-1		0	1
0		1	0
1	1	$0 - 1 \cdot 1 = -1$	$1 - 1 \cdot 0 = 1$
2	5	$1 - 5(-1) = 6$	$0 - 5 \cdot 1 = -5$
3	3	$-1 - 3 \cdot 6 = -19$	$1 - 3(-5) = 16$
4	1	$6 - 1(-19) = 25$	$-5 - 1 \cdot 16 = -21$
5	3	$-19 - 3 \cdot 25 = -94$	$16 - 3(-21) = 79$

From the $i = 4$ line, we see that $158x + 188y = 2$ for $x = 25$ and $y = -21$. These are the same values we got by solving the equations backward in Section 1.3. Computing the $i = 5$ line was not necessary but does provide a check.

Notice that since $r_5 = r_{n+1} = 0$ we have $ax_5 + by_5 = 158(-94) + 188 \cdot 79 = 0$. In general if r_n is the last nonzero remainder we have $ax_{n+1} + by_{n+1} = 0$, so $x_{n+1}/y_{n+1} = -b/a$. ◇

One advantage this method has over that of solving the Euclidean algorithm equations backward is that once a quotient q_i is used in the calculation, it is not needed any longer and can be forgotten. In fact, the calculation of the coefficients x_i and y_i can be done as the Euclidean algorithm equations are derived. This is especially useful in a computer program, since using the previous method would require the computer to remember all the quotients and remainders before it could start to find x and y.

Continued Fractions

In our example, the coefficients x_i and y_i alternated in sign, even if all the quotients q_i were positive. To simplify some later calculations we will introduce some new variables that will turn out to be positive when all the quotients q_i are positive. We define

$$h_i = (-1)^i x_i, \quad k_i = (-1)^{i+1} y_i,$$

for $i = -1, 0, 1, \ldots$. We can use (6.3) to get recursions for these new variables. In particular

$$\begin{aligned} h_i &= (-1)^i (x_{i-2} - q_i x_{i-1}) \\ &= (-1)^{i-2} x_{i-2} + q_i (-1)^{i-1} x_{i-1} \\ &= h_{i-2} + q_i h_{i-1}; \end{aligned}$$

and a similar relation holds for k_i. The reader should check that from the definitions of h_i and k_i and the definitions of x_i and y_i for $i = -1$ and 0 we have $h_{-1} = 0$, $k_{-1} = 1$, $h_0 = 1$, and $k_0 = 0$. Although the numbers q_i have been (and usually will be) integers, the numbers h_i and k_i are defined for any finite sequence of real numbers $q_1, q_2, \ldots, q_{n+1}$.

DEFINITION. Let $q_1, q_2, \ldots, q_{n+1}$ be real numbers. We define h_i and k_i, $i = -1, 0, 1, \ldots, n+1$ by

$$\begin{aligned} &h_{-1} = 0, \quad h_0 = 1, \qquad h_i = h_{i-2} + q_i h_{i-1} \text{ for } i \geq 1, \\ &k_{-1} = 1, \quad k_0 = 0, \qquad k_i = k_{i-2} + q_i k_{i-1} \text{ for } i \geq 1. \end{aligned}$$

Example. Compute h_i and k_i for $i = 1, 2$, and 3 in terms of the numbers q_i.

We have

$$\begin{aligned} h_1 &= h_{-1} + q_1 h_0 = 0 + q_1 \cdot 1 = q_1, \\ k_1 &= k_{-1} + q_1 k_0 = 1 + q_1 \cdot 0 = 1. \end{aligned}$$

In the same way, we compute

$$\begin{aligned} h_2 &= 1 + q_1 q_2, \\ k_2 &= q_2, \\ h_3 &= q_1 + q_3(1 + q_1 q_2), \\ k_3 &= 1 + q_2 q_3. \end{aligned}$$

◇

The ratios h_i/k_i have a curious form. For example,

$$\begin{aligned} \frac{h_1}{k_1} &= \frac{q_1}{1} = q_1, \\ \frac{h_2}{k_2} &= \frac{1 + q_1 q_2}{q_2} = q_1 + \frac{1}{q_2} \quad \text{and} \\ \frac{h_3}{k_3} &= \frac{q_1 + q_3 + q_1 q_2 q_3}{1 + q_2 q_3} \\ &= \frac{q_1(1 + q_2 q_3) + q_3}{1 + q_2 q_3} = q_1 + \frac{q_3}{1 + q_2 q_3} = q_1 + \cfrac{1}{q_2 + \cfrac{1}{q_3}}. \end{aligned}$$

Since the notation is getting messy, we introduce a more compact form.

DEFINITION. finite continued fraction, simple finite continued fraction

By a **finite continued fraction** we mean an expression of the form

$$q_1 + \cfrac{1}{q_2 + \cfrac{1}{q_3 + \cfrac{1}{\ddots + \cfrac{1}{q_r}}}},$$

where the q_i are real numbers, positive for i greater than 1. We denote this by $[q_1, q_2, \ldots, q_r]$. If all the numbers q_i are integers we say the continued fraction is **simple**.

Examples. From the definition

$$\begin{aligned} [-2, 3, 5] &= -2 + \cfrac{1}{3 + \cfrac{1}{5}} \\ &= -2 + \cfrac{1}{\left(\cfrac{16}{5}\right)} = -2 + \frac{5}{16} = \frac{-27}{16}. \end{aligned}$$

Likewise $[1,5,3,1,3] =$

$$1+\cfrac{1}{5+\cfrac{1}{3+\cfrac{1}{1+\cfrac{1}{3}}}} = 1+\cfrac{1}{5+\cfrac{1}{3+\cfrac{3}{4}}} = 1+\cfrac{1}{5+\cfrac{4}{15}} = 1+\frac{15}{79} = \frac{94}{79}.$$

$\diamondsuit$

Usually we will be interested in only simple continued fractions, as in our examples. Requiring $q_i > 0$ for $i > 1$ means that we do not have to worry about dividing by zero in our expressions. Notice that the definition implies that

$$[q_1, q_2, \ldots, q_r] = \left[q_1, q_2, \ldots, q_{r-1} + \frac{1}{q_r}\right]. \tag{6.4}$$

Now we have a compact way of expressing the pattern we discovered for the ratios h_i/k_i, namely,

$$\begin{aligned}\frac{h_1}{k_1} &= [q_1],\\ \frac{h_2}{k_2} &= [q_1, q_2], \quad \text{and}\\ \frac{h_3}{k_3} &= [q_1, q_2, q_3].\end{aligned}$$

Theorem 6.8. *Let $q_1, q_2, \ldots, q_r$ be real numbers, with $q_i > 0$ for $i > 1$, and let h_i and k_i be defined as above. Then*

$$[q_1, q_2, \ldots, q_r] = \frac{h_r}{k_r}.$$

Proof. The proof will be by induction on r. We already know the theorem is true for $r = 1, 2$, and 3. Suppose we know the theorem to be true for $r = n$. Let $q_1, q_2, \ldots, q_{n+1}$ be given, and let h_i and k_i be defined as usual for $i = -1, 0, 1, \ldots, n+1$.

We define another sequence of n real numbers $Q_1, Q_2, \ldots, Q_n$ by

$$Q_1 = q_1, \quad Q_2 = q_2, \quad \ldots, \quad Q_{n-1} = q_{n-1}, \quad \text{and} \quad Q_n = q_n + \frac{1}{q_{n+1}}.$$

Let the number H_i and K_i be defined for $i = -1, 0, \ldots, n$ just as the numbers h_i and k_i, but in terms of the numbers Q_i instead of q_i. Note that $H_i = h_i$ and

$K_i = k_i$ for $i < n$ since $Q_i = q_i$ for $i < n$. Then

$$\begin{aligned}
[q_1, q_2, \dots, q_{n+1}] &= \left[q_1, \dots, q_{n-1}, q_n + \frac{1}{q_{n+1}}\right] \text{ (by (6.4))} \\
&= [Q_1, \dots, Q_n] = \frac{H_n}{K_n} \text{ (by the induction hypothesis)} \\
&= \frac{H_{n-2} + Q_n H_{n-1}}{K_{n-2} + Q_n K_{n-1}} = \frac{h_{n-2} + \left(q_n + \frac{1}{q_{n+1}}\right) h_{n-1}}{k_{n-2} + \left(q_n + \frac{1}{q_{n+1}}\right) k_{n-1}} \\
&= \frac{h_{n-2} + q_n h_{n-1} + \frac{h_{n-1}}{q_{n+1}}}{k_{n-2} + q_n k_{n-1} + \frac{k_{n-1}}{q_{n+1}}} \\
&= \frac{h_n + \frac{h_{n-1}}{q_{n+1}}}{k_n + \frac{k_{n-1}}{q_{n+1}}} = \frac{h_{n-1} + q_{n+1} h_n}{k_{n-1} + q_{n+1} k_n} = \frac{h_{n+1}}{k_{n+1}}.
\end{aligned}$$

□

Example. Compute $[-2, 3, 5]$ by means of Theorem 6.8.

We have $[-2, 3, 5] = h_3/k_3$, where the computation of the integers h_i and k_i is displayed in the following table.

i	q_i	$h_i = h_{i-2} + q_i h_{i-1}$	$k_i = k_{i-2} + q_i k_{i-1}$
-1		0	1
0		1	0
1	-2	$0 + (-2)1 = -2$	$1 + (-2)0 = 1$
2	3	$1 + 3(-2) = -5$	$0 + 3 \cdot 1 = 3$
3	5	$-2 + 5(-5) = -27$	$1 + 5 \cdot 3 = 16$

We see that $[-2, 3, 5] = -27/16$. ◇

Theorem 6.9. *Any simple continued fraction is equal to a rational number. Conversely, any nonintegral rational number can be expressed as a simple finite continued fraction $[q_1, q_2, \dots, q_{n+1}]$, where $q_{n+1} > 1$.*

Proof. The first statement follows from Theorem 6.8 since if $q_1, \dots, q_r$ are all integers, then so are all the numbers h_i and k_i, and so h_r/k_r is rational. (It is also obvious from the definition of a simple finite continued fraction.)

Conversely, any nonintegral rational number can be written as b/a with $a > 1$ and $(a, b) = 1$. Apply the Euclidean algorithm as usual, getting quotients q_i and remainders r_i. Let r_n be the last nonzero remainder. Then by Theorem 6.7

$$0 = r_{n+1} = ax_n + by_n,$$

so

$$\frac{b}{a} = \frac{-x_{n+1}}{y_{n+1}} = \frac{h_{n+1}}{k_{n+1}}.$$

This is $[q_1, q_2, \dots, q_{n+1}]$ by Theorem 6.8.

The last equation in the Euclidean algorithm is

$$r_{n-1} = r_n q_{n+1} + 0,$$

where if $n = 1$ we interpret r_0 to be a. Since $1 = (a, b) = r_n < r_{n-1}$, this equation implies that $q_{n+1} > 1$. □

Note that the integer n also has the continued fraction representation $[n]$; we just cannot be sure that $n > 1$.

Problems for Section 6.2.

A

In the first four problems, assume the given triple is q_1, q_2, q_3. Find x_i and y_i for $i = 1, 2, 3$.

1. 6, 1, 2
2. 4, 9, 3
3. −2, 5, 7
4. −3, 1, 10

In the next six problems, find x and y such that $ax + by = (a, b)$ by the method of Theorem 6.7.

5. $a = 51, b = 71$
6. $a = 92, b = 300$
7. $a = 731, b = 45$
8. $a = 666, b = 101$
9. $a = 402, b = -1005$
10. $a = 671, b = -616$

In the next four problems, assume the given triple is q_1, q_2, q_3. Find h_i and k_i for $i = 1, 2, 3$.

11. 2, 3, 4
12. 5, 4, 3
13. −4, 1, 6
14. 0, 5, 7

In the next four problems, evaluate the given continued fraction as a rational number by the method of the examples after the definition of a finite continued fraction. Then confirm your work by using Theorem 6.8.

15. $[2, 4, 1, 5]$
16. $[6, 1, 1]$
17. $[0, 3, 7, 5, 2]$
18. $[-5, 2, 1, 2]$

In the next six problems, write each rational number b/a as a continued fraction $[q_1, q_2, \dots, q_{n+1}]$ by applying the Euclidean algorithm to a and b.

19. 73/41
20. 41/73
21. −45/14
22. −34/105
23. 3.14
24. 3.1416

B

Define $R_i = h_i/k_i$. In each of the next four problems, let the number given be b/a. Let $q_1, q_2, \dots, q_{n+1}$ be the quotients when the Euclidean algorithm is applied to a and b. Find R_i for $i = 1$ to $n+1$, and label these points on the real axis, along with b/a.

25. 71/15 26. 101/43 27. 3.14 28. 2.72

29. Find expressions for h_4 and k_4 in terms of the numbers q_i.
30. Show that if $q_r > 1$, then $[q_1, q_2, \ldots, q_r] = [q_1, q_2, \ldots, q_{r-1}, q_r - 1, 1]$.
31. Prove that k_i does not depend on q_1.
32. Prove that if $[q_1, \ldots, q_r]$ is a finite continued fraction, then $k_i > 0$ for $i = 1, 2, \ldots, r$.
33. Prove that $h_i = (1 + q_i q_{i-1})h_{i-2} + q_i h_{i-3}$ for $i > 1$.
34. Show that $k_i > k_{i-1}$ for $i > 2$ if $q_i > 0$ for $i > 1$.

C

If the rational number b/a is expressed as the simple continued fraction $[q_1, \ldots, q_{n+1}]$ with q_{n+1} greater than 1, then it also has the representation $[q_1, \ldots, q_{n+1} - 1, 1]$, as is shown in problem 30. It turns out that every rational number has exactly these two representations, and no more. This is proved in the next four problems.

35. Show that each rational number has at least two distinct representations as a simple finite continued fraction.
36. Suppose $R = [a_1, \ldots, a_m] = [b_1, \ldots, b_n]$, where these are simple finite continued fractions with $m > 1$, $n > 1$, $a_m > 1$, and $b_n > 1$. Show that $a_1 = b_1$ and $[a_2, \ldots, a_m] = [b_2, \ldots, b_n]$. (**Hint:** Show $a_1 = \lfloor R \rfloor$, where this is the greatest integer function defined in Section 6.1.)
37. Show that under the hypotheses of the last problem $m = n$ and $a_i = b_i$ for $i = 1, \ldots, m$.
38. Show that any rational number can be expressed as a finite simple continued fraction in exactly two ways.

6.3 Infinite Continued Fractions

In the last section, we saw that every finite sequence of integers $q_1, \ldots, q_r$ with $q_i > 0$ for $i > 1$ corresponds to a rational number expressed by the continued fraction $[q_1, \ldots, q_r]$, and that each rational number has such a representation. In this section, we will try to make sense out of an expression of the form $[q_1, q_2, \ldots]$, where the numbers q_i form an infinite sequence. To understand this section fully the reader should be familiar with the meaning of a statement of the form

$$\lim_{n \to \infty} R_n = L,$$

where $R_1, R_2, \ldots$ is an infinite sequence of real numbers, but a useful, if somewhat more vague, understanding of infinite continued fractions is possible even without a knowledge of limits.

It turns out that there is a one-to-one correspondence between infinite continued fractions and the irrational numbers, so our analysis will consist of two parts: (1) showing how each infinite sequence $q_1, q_2, \ldots$ defines an irrational number, and (2) showing how each irrational number produces an infinite sequence $q_1, q_2, \ldots$.

Making Sense out of $[q_1, q_2, \ldots]$

A natural definition for $[q_1, q_2, \ldots]$ would be as the limit of $[q_1, \ldots, q_n]$ as $n \to \infty$, but, of course, it is not obvious whether the limit exists or not. The key to our analysis will be the result from the previous section that $[q_1, \ldots, q_n] = h_n/k_n$, where h_n and k_n are as defined there. We need to know more about the relations among these variables.

Since some of our results will apply to finite continued fractions as well as infinite ones, we will assume for the time being that $q_1, q_2, \ldots$ is a finite or infinite sequence of integers, with $q_i > 0$ for $i > 1$ (whenever q_i is defined, of course).

Lemma 6.10. *We have* $1 = k_1 \leq k_2 < k_3 < \ldots$.

Proof. In Section 6.2, we computed that $k_1 = 1$ and $k_2 = q_2$, which is at least 1 by our assumption about the q_i. The equation $k_i = k_{i-2} + q_i k_{i-1}$ then shows that $k_i \geq k_{i-2} + k_{i-1} > k_{i-1}$ for $i > 2$. □

Consider the equations

$$\begin{aligned} h_i &= h_{i-2} + q_i h_{i-1}, \\ k_i &= k_{i-2} + q_i k_{i-1}. \end{aligned}$$

Suppose we eliminate q_i by multiplying the first equation by k_{i-1}, the second by h_{i-1}, and subtracting. We get

$$k_{i-1}h_i - h_{i-1}k_i = k_{i-1}h_{i-2} - h_{i-1}k_{i-2} = -(k_{i-2}h_{i-1} - h_{i-2}k_{i-1}),$$

which is just the negative of what we started with, but with all subscripts decreased by 1. Since $k_{-1}h_0 - h_{-1}k_0 = 1 \cdot 1 - 0 \cdot 0 = 1$, we have $k_0 h_1 - h_0 k_1 = -1$, $k_1 h_2 - h_1 k_2 = 1$, etc.

Lemma 6.11. *We have*

$$k_{i-1}h_i - h_{i-1}k_i = (-1)^i \quad \text{for} \quad i = 0, 1, \ldots .$$

In consequence, h_i and k_i are relatively prime for all i.

The second statement of this lemma, which follows from Theorem 1.10, says that the fraction h_i/k_i is in lowest terms. We will denote this fraction by R_i. By Lemma 6.10, R_i makes sense for $i > 0$. Dividing the equation of Lemma 6.11 by $k_i k_{i-1}$ yields

$$R_i - R_{i-1} = \frac{(-1)^i}{k_i k_{i-1}}. \tag{6.5}$$

Likewise

$$R_{i+1} - R_i = \frac{(-1)^{i+1}}{k_{i+1}k_i},$$

and adding this to (6.5) gives us

$$R_{i+1} - R_{i-1} = \frac{(-1)^i \left(\frac{1}{k_{i-1}} - \frac{1}{k_{i+1}} \right)}{k_i}.$$

By Lemma 6.10 we have

$$k_{i+1} > k_{i-1} \quad \text{for} \quad i > 1,$$

so $R_{i+1} - R_{i-1}$ is positive or negative according as i is even or odd. Letting $i = 2, 3, \ldots$ gives $R_3 - R_1 > 0$, $R_4 - R_2 < 0$, $R_5 - R_3 > 0$, etc. This proves the following result.

Lemma 6.12. *We have*

$$R_1 < R_3 < R_5 < \cdots \quad \text{and} \quad R_2 > R_4 > R_6 > \cdots .$$

Example. Consider the finite continued fraction $[1, 2, 3, 1]$. Compute the ratios R_i and graph them on the real axis.

The numbers h_i, k_i, and R_i are computed in the following table.

i	-1	0	1	2	3	4
q_i			1	2	3	1
h_i	0	1	1	3	10	13
k_i	1	0	0	2	7	9
R_i			1	$3/2 = 1.5$	$10/7 \approx 1.43$	$13/9 \approx 1.44$

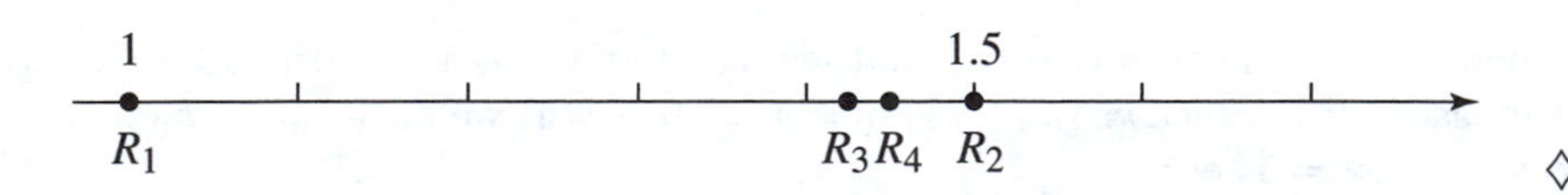

◇

Lemma 6.13. *Suppose i and j are positive integers, with i even and j odd. Then $R_i > R_j$.*

Proof. Let n be the larger of i and j. From (6.5) and Lemma 6.10, the quantity $R_n - R_{n-1}$ is positive or negative according as n is i or j. Then by Lemma 6.12 we have in the first case

$$R_i > R_{i-1} \geq R_j,$$

and in the second case

$$R_j < R_{j-1} \leq R_i.$$

□

Theorem 6.14. *Let $q_1, q_2, \ldots$ be an infinite sequence of integers, with $q_i > 0$ for $i > 1$. Then*

$$\lim_{r \to \infty} [q_1, q_2, \ldots, q_r]$$

exists, and is an irrational number.

Proof. By Lemmas 6.12 and 6.13 the numbers $R_1, R_3, \ldots$ form an increasing sequence of real numbers that is bounded above by R_2, and so has a limit. Likewise, $R_2, R_4, \ldots$ is a decreasing sequence bounded below by R_1, and so also has a limit. Since Lemma 6.10 implies the integers k_i go to infinity, by (6.5) these two limits must be the same, say the real number R. This proves the first statement of the theorem. Note that

$$R_1 < R_3 < \cdots \leq R \leq \cdots < R_4 < R_2, \tag{6.6}$$

so R is not equal to any quotient R_i.

Now we prove that R is irrational. Suppose $R = b/a$, with $a > 0$. From (6.6) b/a is strictly between R_i and R_{i-1}, and so

$$0 < \left| \frac{b}{a} - R_{i-1} \right| < |R_i - R_{i-1}|.$$

The quantity on the right equals $1/k_i k_{i=1}$ by (6.5). If we multiply through by ak_{i-1}, we get

$$0 < |bk_{i-1} - ah_{i-1}| < \frac{a}{k_i},$$

where we have used the fact that $R_{i-1} = h_{i-1}/k_{i-1}$.

Since we can make k_i as large as we please by taking i large, we can make the quantity on the right less than 1. But the quantity inside the absolute value signs is an integer. We have an integer strictly between 0 and 1, which is impossible. This contradiction shows that R must be irrational. □

DEFINITION. infinite continued fraction, convergent

Let $q_1, q_2, \ldots$ be an infinite sequence of integers, positive except perhaps for q_1. We define

$$[q_1, q_2, \ldots] = \lim_{n \to \infty} [q_1, q_2, \ldots, q_n].$$

This is called an **infinite continued fraction**. The numbers $[q_1, q_2, \ldots, q_n] = R_n = h_n/k_n$, $n = 1, 2, \ldots$, are said to be the **convergents** of the infinite continued fraction.

Given a sequence $q_1, q_2, \ldots$ it is possible to approximate the corresponding irrational number $[q_1, q_2, \ldots]$ as closely as one wants by computing the numbers R_i. In fact, by (6.5) and (6.6) the error in using R_{i-1} to approximate R is less than $1/k_i k_{i-1}$. Usually it is not possible to name R other than by the continued fraction, however. An exception occurs when the sequence $q_1, q_2, \ldots$ repeats.

Example. What irrational number does $[1, 2, 1, 2, 1, 2, \ldots]$ represent?

Call the number R. Note that

$$R_{2n} = 1 + \cfrac{1}{2 + \cfrac{1}{R_{2n-2}}}.$$

(Write out $R_2 = [1, 2]$, $R_4 = [1, 2, 1, 2]$, and $R_6 = [1, 2, 1, 2, 1, 2]$ to get the idea.) If we let n go to infinity, we see that

$$R = 1 + \cfrac{1}{2 + \cfrac{1}{R}}.$$

This leads to the equation

$$2R^2 - 2R - 1 = 0,$$

which has the solutions $R = (1 \pm \sqrt{3})/2$. Since R is positive taking the plus sign gives its value. ◊

Finding the Continued Fraction for an Irrational Number

Now we will consider the converse problem of finding the infinite continued fraction corresponding to a given irrational number. Let the number be S. We need a sequence $q_1, q_2, \ldots$ such that $S = [q_1, q_2, \ldots]$.

If S were rational, say b/a, we would find the quotients q_i by applying the Euclidean algorithm to a and b, getting

$$\begin{aligned} b &= a q_1 + r_1, \quad 0 < r_1 < a, \\ a &= r_1 q_2 + r_2, \quad 0 < r_2 < r_1, \end{aligned}$$

and so on. If we divide the first equation and inequality by a we get

$$\frac{b}{a} = q_1 + \frac{r_1}{a}, \quad 0 < \frac{r_1}{a} < 1.$$

We see that q_1 is the integer part and r_1/a is the fractional part of b/a, that is

$$q_1 = \left\lfloor \frac{b}{a} \right\rfloor, \quad \frac{r_1}{a} = \frac{b}{a} - \left\lfloor \frac{b}{a} \right\rfloor.$$

In the same way, dividing the second equation and inequality by r_1 gives

$$\frac{a}{r_1} = q_2 + \frac{r_2}{r_1}, \quad 0 < \frac{r_2}{r_1} < 1.$$

Thus

$$q_2 = \left\lfloor \frac{a}{r_1} \right\rfloor = \left\lfloor \frac{1}{r_1/a} \right\rfloor = \left\lfloor \frac{1}{b/a - \lfloor b/a \rfloor} \right\rfloor.$$

We need some notation to simplify what is going on here. Let $S_1 = b/a$. Then

$$q_1 = \lfloor S_1 \rfloor.$$

Also if we define S_2 to be $1/(S_1 - q_1)$, then

$$q_2 = \lfloor S_2 \rfloor.$$

The generalization of this process gives us a method for finding the continued fraction expression for an irrational number.

Theorem 6.15. *Let S be a real number. Let $S_1 = S$ and $q_1 = \lfloor S_1 \rfloor$. Suppose S_i and q_i have been defined for $i = 1, 2, \ldots, r$. If S_i does not equal q_i, then define S_{i+1} and q_{i+1} by*

$$S_{i+1} = \frac{1}{S_i - q_i}, \quad q_{i+1} = \lfloor S_{i+1} \rfloor.$$

Then if S is rational this process stops after the definition of some q_r, and we have

$$S = [q_1, q_2, \ldots, q_r],$$

while if S is irrational an infinite sequence $q_1, q_2, \ldots$ is defined and

$$S = [q_1, q_2, \ldots].$$

Before proving this theorem we will do an example of the computation of the numbers S_i and q_i. We will take $S = \pi$, which is irrational (although this fact will not be proved in this book). Then $S_1 = S = 3.1415926535897\ldots$, and so $q_1 = \lfloor 3.14\ldots \rfloor = 3$. (The first 14 digits of π may be remembered from the number of letters in the words "how I want a drink, alcoholic of course, after the heavy integral theoretic lecture.") Then $S_2 = 1/(\pi - 3) = 7.0625\ldots$ and $q_2 = 7$. Likewise $S_3 = 1/.0625\ldots = 15.9965\ldots$ and so $q_3 = 15$. So far we have $\pi = [3, 7, 15, \ldots]$.

In practice, if we start with a value of π from a calculator or table of constants, which is only an approximation, the numbers q_i we compute will eventually diverge from the true values, how soon depending on the accuracy of the original approximation. Using a calculator that gives π to 9 digits to the right of the decimal point, for example, we compute the continued fraction $[3, 7, 15, 1, 293, 10, 3, 8, 2, 1, \ldots]$ for π. The correct infinite continued fraction, however, is known to start out $[3, 7, 15, 1, 292, 1, 1, 1, 2, 1, \ldots]$. (If there is any pattern in this expansion, no one has ever figured it out.)

The values of R_n turn out to be good rational approximations to the irrational number S. What "good" means will not be made precise here; but roughly, we want R_n close to S without having too large a denominator k_n.

In this example, we have $R_1 = [3] = 3$, a number claimed to be π by Leeds K. Field in his 1953 book *Mathematics, Minus and Plus.* Likewise $R_2 = [3, 7] = 22/7$, which is the usual rational approximation, and correct within about 4/100 of 1%. Another amazingly good approximation to π is $R_4 = [3, 7, 15, 1] = 355/113$, with an error of less than one part in 10 million.

Proof of Theorem 6.15. If S is rational, it is not hard to see that the integers q_i we get are exactly the quotients q_i of the Euclidean algorithm, so we will concentrate on the case when S is irrational.

Since $S = S_1$ is irrational, it cannot equal the integer q_1, and so $S_2 = 1/(S_1 - q_1)$ is defined. In fact, S_2 must be irrational too, since otherwise $S_1 = q_1 + 1/S_2$ would be rational. A continuation of this argument shows that all the numbers S_i are defined and irrational, so that the sequence $q_1, q_2, \ldots$ is infinite.

Note that since $S_i - q_i$ is the fractional part of S_i, we have $0 < S_i - q_i < 1$, and so $S_{i+1} = 1/(S_i - q_i) > 1$. Thus $q_{i+1} = \lfloor S_{i+1} \rfloor \geq 1$ for $i > 0$.

Then by Theorem 6.14 we have $[q_1, q_2, \ldots] = R$ for some irrational number R. Let the numbers h_i, k_i, and R_i be defined as before, so that R is the limit of the ratios R_i. Note that

$$S_2 = \frac{1}{S_1 - q_1}, \quad \text{so} \quad S = S_1 = q_1 + \frac{1}{S_2},$$

and, in general,

$$S_i = q_i + \frac{1}{S_{i+1}}.$$

Then

$$S = S_1 = q_1 + \frac{1}{S_2} = q_1 + \cfrac{1}{q_2 + \cfrac{1}{S_3}} = \cdots = [q_1, \ldots, q_{n-1}, S_n].$$

By Theorem 6.8 the last expression equals

$$\frac{h_{n-2}+S_n h_{n-1}}{k_{n-2}+S_n k_{n-1}}.$$

Thus

$$\begin{aligned} S-R_{n-1} &= \frac{h_{n-2}+S_n h_{n-1}}{k_{n-2}+S_n k_{n-1}}-\frac{h_{n-1}}{k_{n-1}} \\ &= \frac{k_{n-1}(h_{n-2}+S_n h_{n-1})-(k_{n-2}+S_n k_{n-1})h_{n-1}}{k_{n-1}(k_{n-2}+S_n k_{n-1})} \\ &= \frac{-(k_{n-2}h_{n-1}-h_{n-2}k_{n-1})}{k_{n-1}(k_{n-2}+S_n k_{n-1})}=\frac{(-1)^n}{k_{n-1}(k_{n-2}+S_n k_{n-1})} \end{aligned}$$

by Lemma 6.11. Now the k_i go to infinity and $S_n > 0$ for $n > 1$ so the last quantity goes to 0 as n goes to infinity. We see

$$0=\lim_{n\to\infty} S-R_{n-1}=S-R.$$

□

Expansions of Quadratic Irrationalities

If the irrational number S is of the form $A\sqrt{d}+B$, where A and B are rational numbers and d is a positive integer, then we can find its (infinite) continued fraction expansion explicitly. In fact, although we will not prove it here, it turns out that such numbers are exactly those having expansions that repeat past some point.

DEFINITION. periodic continued fraction

An infinite continued fraction $[q_1, q_2, \ldots]$ is said to be **periodic** if it repeats from some point on, that is, if there exist positive integers m and r such that $q_n = q_{n+r}$ for $n > m$. An example of a periodic expansion is

$$[-4,5,1,2,7,1,3,7,1,3,7,1,3,\ldots],$$

where the pattern 7-1-3 keeps repeating. (Here $m = 4$ and $r = 3$.) We denote this continued fraction by

$$[-4,5,1,2,\overline{7,1,3}],$$

where the bar indicates the group of q_i under it repeats indefinitely. As another example of this notation, from the example after Theorem 6.14 we have

$$(1+\sqrt{3})/2=[1,2,\overline{1,2}]=[1,\overline{2,1}]=[\overline{1,2}].$$

Example. Find the continued fraction expansion of $S=\frac{1}{6}\sqrt{21}+\frac{7}{2}$.

Let $S_1 = S = (\sqrt{21} + 21)/6$. This number is about 4.3, so $q_1 = \lfloor S_1 \rfloor = 4$. Then

$$\begin{aligned} S_2 &= \left(\frac{\sqrt{21}+21}{6} - 4\right)^{-1} = \left(\frac{\sqrt{21}-3}{6}\right)^{-1} \\ &= \frac{6}{\sqrt{21}-3} = \frac{6(\sqrt{21}+3)}{(\sqrt{21}-3)(\sqrt{21}+3)} = \frac{6(\sqrt{21}+3)}{21-9} = \frac{\sqrt{21}+3}{2}. \end{aligned}$$

This number is about 3.8, so $q_2 = 3$. Then

$$\begin{aligned} S_3 &= \left(\frac{\sqrt{21}+3}{2} - 3\right)^{-1} = \left(\frac{\sqrt{21}-3}{2}\right)^{-1} \\ &= \frac{2}{\sqrt{21}-3} = \frac{2(\sqrt{21}+3)}{21-9} = \frac{\sqrt{21}+3}{6}. \end{aligned}$$

Since S_3 is about 1.6, we have $q_3 = 1$. Then

$$S_4 = \left(\frac{\sqrt{21}+3}{6} - 1\right)^{-1} = \left(\frac{\sqrt{21}-3}{6}\right)^{-1}.$$

But this is the same as S_2. Thus $q_4 = q_2 = 3$, $q_5 = q_3 = 1$, and the expansion repeats from now on. We see

$$\frac{\sqrt{21}+21}{6} = [4, 3, 1, 3, 1, \ldots] = [4, \overline{3, 1}].$$

◇

Example. Find the irrational number represented by $S = [-1, 1, \overline{1, 2, 3}]$.
We have

$$S_1 = -1 + \cfrac{1}{1 + \cfrac{1}{S_3}}, \quad \text{where} \quad S_3 = [\overline{1, 2, 3}] = 1 + \cfrac{1}{2 + \cfrac{1}{3 + \cfrac{1}{S_3}}}.$$

Simplifying the last equation leads to

$$7S_3^2 - 8S_3 - 3 = 0,$$

which has the solutions $(4 \pm \sqrt{37})/7$. Since $S_3 > 0$, we choose the plus sign.

Finally by plugging this value of S_3 into the first equation of this example, simplifying, and rationalizing the denominator, we get

$$S_1 = S = \frac{\sqrt{37} - 11}{12}.$$

Problems for Section 6.3.

A

In the first eight problems, find the irrational number corresponding to the given periodic continued fraction.

1. $[\overline{1}]$
2. $[\overline{2}]$
3. $[\overline{2,1}]$
4. $[\overline{1,3}]$
5. $[1,\overline{3}]$
6. $[-2,\overline{1}]$
7. $[-1,3,\overline{2,3}]$
8. $[-3,1,\overline{1,4}]$

In the next eight problems, find the periodic continued fraction for the given irrational number.

9. $(\sqrt{5}-1)/2$
10. $\sqrt{2}$
11. $(5-\sqrt{13})/6$
12. $(17-\sqrt{3})/13$
13. $2\sqrt{3}$
14. $1-2/\sqrt{3}$
15. $(5+\sqrt{37})/4$
16. $(\sqrt{37}+5)/3$

In the next four problems, find the first four terms of the continued fraction expansion of the given number.

B

17. $\pi/4$
18. $2^{1/3}$
19. $\sin 10^\circ$
20. e
21. Show that if m is a positive integer, then $[\overline{m}] = (m+\sqrt{m^2+4})/2$.
22. Show that if a and b are integers > 0, then $[\overline{a,b}] = (ab+\sqrt{a^2b^2+4ab})/2b$.
23. Let A, B, A', and B' be rational numbers and d a positive integer, not a perfect square. Show that if $A\sqrt{d}+B = A'\sqrt{d}+B'$, then $A=A'$ and $B=B'$.
24. Let R be irrational. Show that R has the form $A\sqrt{d}+B$, where A and B are rational and d is a positive integer if and only if R is a solution of an equation $ax^2+bx+c=0$ with a, b, and c integers.
25. Show that the continued fraction expansion of an irrational number is unique.
26. Let m be a positive integer. Show that $\sqrt{m^2+1} = [m,\overline{2m}]$.
27. From the proof of Theorem 6.14, we see that if $[q_1,q_2,\ldots] = R$, then $|R-R_{i-1}| < 1/k_ik_{i-1}$. Assuming $\pi = [3,7,15,\ldots]$, how close does this say π is to $22/7$?
28. Assuming $\pi = [3,7,15,1,292,\ldots]$, how close must it be to $355/113$? (See the previous problem.)
29. Let $q_i = 1$ for all $i > 0$. Show that $f_i = k_i = h_{i-1}$ for $i > 0$, where f_i is the ith Fibonacci number.

C

In the next three problems, assume that $R = [\overline{q_1,q_2,\ldots,q_n}]$, where the q_i are positive integers.

30. Show that $R = [q_1,q_2,\ldots,q_n,R]$.
31. Show that if the numbers h_i and k_i are defined as usual, then $R = (h_{n-1}+Rh_n)/(k_{n-1}+Rk_n)$, where h_{n-1}, k_{n-1}, h_n, and k_n are integers.
32. Show that R satisfies a quadratic equation with integer coefficients.

6.4 Decimal Representation

In our study of the continued fraction expansions of real numbers, we found a tidy matching of finite expansions with rational and infinite expansions with irrational numbers. A more familiar way of distinguishing rational from irrational numbers has to do with whether their decimal expansions repeat or not.

DEFINITION. periodic decimal, period, terminating decimal

Consider the decimal representation

$$\pm e_1 e_2 \cdots e_m . d_1 d_2 \cdots .$$

We say that this expansion is **periodic** if there exist positive integers k and j such that $d_i = d_{i+j}$ for $i > k$. If j is the least such positive integer, then we call j the **period** of the expansion, and denote the expansion by

$$\pm e_1 e_2 \cdots e_m . d_1 d_2 \ldots d_k \overline{d_{k+1} \cdots d_{k+j}}.$$

For example, $2/3 = .6666\ldots = .\overline{6}$, and the period of this expansion is 1. Likewise $.165214214214\ldots = .165\overline{214}$, with period 3. If $j = 1$ and $d_i = 0$ for $i > k$, then we say the expansion is **terminating**. An example of a terminating expansion is $9/4 = 2.25 = 2.25\overline{0}$.

Since any real number can be written in the form $\pm(n + X)$, where n is a nonnegative integer and $0 \leq X < 1$, we will concentrate on the expansions of numbers X between 0 and 1. We will assume certain facts:

1. Each real number X, $0 \leq X < 1$, has a decimal expansion.

2. This expansion is unique except that the terminating decimal $.d_1 d_2 \cdots d_k \overline{0}$, with $d_k > 0$, also has the expansion $.d_1 d_2 \cdots d_{k-1}(d_k - 1)\overline{9}$. For example, $.135 = .135\overline{0} = .134\overline{9} = .134999\ldots.$

3. A decimal expansion for the rational number b/a can be found by dividing b by a using the usual long division algorithm taught in elementary school.

4. Decimals can be added and subtracted in the usual way, and multiplied or divided by powers of 10 by moving the decimal point appropriately.

Theorem 6.16. *A real number is rational if and only if it has a periodic decimal expansion.*

Proof. If Y is any real number, we can write $Y = \pm(n + X)$, where $0 \leq X < 1$, and n is a nonnegative integer. Since Y has a periodic expansion if and only if X does, and is rational if and only if X is, it suffices to prove the theorem for numbers between 0 and 1.

We will start by assuming that X has the periodic representation

$$.d_1 d_2 \cdots d_k \overline{d_{k+1} \cdots d_{k+j}}.$$

Then

$$10^{k+j}X - 10^k X = d_1 \cdots d_{k+j}.\overline{d_{k+1} \cdots d_{k+j}} - d_1 \cdots d_k.\overline{d_{k+1} \cdots d_{k+j}},$$

which is an integer, say s. Then X is the rational number $s/(10^{k+j} - 10^k)$.

Conversely let X be the rational number b/a, where $0 \le b < a$ and $(a, b) = 1$. We can find a decimal expansion for X by the usual algorithm, which we illustrate below for $b = 7$ and $a = 22$.

```
         0 . 3 1 8 1 8 1 8 ...
    22 ) 7 . 0 0 0 0 0 0 0 ...
         6   6
         -------
             4 0
               2 2
             -----
             1 8 0
             1 7 6
             -------
                 4 0
                 2 2
                 -----
                 1 8 0
                   . . .
```

The boldface numbers are the remainders at each stage of the algorithm, and each is nonnegative and less than the divisor, 22. In general, each remainder r satisfies $0 \le r < a$. Since there are only finitely many possible remainders, eventually one must repeat, producing a periodic expansion. In the example, the remainder 4 repeats first, giving the periodic expansion $7/22 = .3\overline{18}$. □

Example. Express $X = .37\overline{216}$ as the quotient of two integers.

We follow the proof of Theorem 6.16 with $k = 2$ and $j = 3$, noting that

$$10^5 X - 10^2 X = 37{,}216.\overline{216} - 37.\overline{216} = 37{,}179.$$

Then $X = 37{,}179/(10^5 - 10^2) = 37{,}179/99{,}900 = 1377/3700$. This can be checked by dividing out the latter fraction. ◇

The Period of a Rational Number

The long division algorithm illustrated in the proof of the last theorem deserves closer study. If the divisor is a, at each stage of the algorithm we get a remainder r_i satisfying $0 \le r_i < a$. Then r_i is multiplied by 10, and $10r_i$ is divided by a to get a new remainder r_{i+1}. We have

$$10r_i = q_{i+1}a + r_{i+1}, \quad 0 \le r_{i+1} < a,$$

where q_{i+1} is the next digit of the quotient. Continuing the process, the next quotient digit q_{i+2} and remainder r_{i+2} satisfy

$$10r_{i+1} = q_{i+2}a + r_{i+2}, \quad 0 \le r_{i+2} < a.$$

Notice that from these equations,

$$r_{i+1} \equiv 10r_i \pmod{a} \quad \text{and} \quad r_{i+2} \equiv 10r_{i+1} \pmod{a}.$$

Substituting the first congruence into the second yields

$$r_{i+2} \equiv 10^2 r_i \pmod{a}.$$

The reader should check that in the same way if r_{i+3} is the next remainder, then

$$r_{i+3} \equiv 10^3 r_i \pmod{a}.$$

In general, if $r_1, r_2, \ldots$ are the remainders produced when b is divided by a using the long division algorithm, where $0 \leq b < a$ and $(a, b) = 1$, then

$$r_{i+j} \equiv 10^j r_i \pmod{a}$$

for all positive integers i and j. In particular, when a remainder repeats, say when $r_{i+j} = r_i$, we have

$$r_i \equiv 10^j r_i \pmod{a}.$$

It would be nice to cancel the r_i from both sides of the congruence, but in order to do this we need to know that $(r_i, a) = 1$.

Let us see how the remainders r_i start out. Since we are assuming that $b < a$, the first step in the long division algorithm is to divide $10b$ by a, getting a quotient digit q_1 (which may be 0) and remainder r_1, satisfying

$$10b = q_1 a + r_1.$$

A good start would be to show that $(r_1, a) = 1$. Suppose some prime p divides both r and a. Then from the last displayed equation, p divides $10b$. Since we are assuming that a and b are relatively prime we cannot have that p divides b, but it could divide 10. (In fact, this happens in the example of the last theorem where $a = 22$; each remainder there is divisible by 2 also.)

Evidently if we want r_1 and a to be relatively prime, we will have to assume that no prime dividing a also divides 10; that is, we must assume that $(a, 10) = 1$. Then the above argument leads to a contradiction, forcing us to the conclusion that $(a, r_1) = 1$.

That all the remainders r_i are relatively prime to a now follows in the same way. For example, we have

$$10r_1 = q_2 a + r_2,$$

and so any prime dividing a and r_2 would have to divide either 10 or r_1, which we know cannot happen. Clearly the argument may be continued indefinitely.

It may be a good idea to recap what we have figured out so far.

Lemma 6.17. *Let b be divided by a by the long division algorithm, where $0 \leq b < a$, and $(a, b) = 1$. Then the remainders r_i satisfy*

$$r_{i+j} \equiv 10^j r_i \pmod{a}$$

for all positive integers i and j. If $(a, 10) = 1$, then all the remainders are relatively prime to a.

Recall that in the first theorem of this section we saw that the digits in the decimal expansion of b/a repeat whenever a remainder repeats itself. Suppose that j is the period of the expansion of b/a, so that for some i we have $r_i = r_{i+j}$. Then

$$r_i \equiv 10^j r_i \pmod{a}.$$

If $(a, 10) = 1$, we know that r_i is relatively prime to a and so can be canceled by the cancellation theorem. This gives

$$10^j \equiv 1 \pmod{a}.$$

We claim that 10^j is the smallest positive power of 10 congruent to 1 modulo a. For suppose $10^k \equiv 1 \pmod{a}$, with $0 < k < j$. Then

$$r_{i+k} \equiv 10^k r_i \equiv r_i \pmod{a},$$

and so $r_i = r_{i+k}$. But then the decimal expansion of b/a repeats every k digits, contradicting the assumption that j is the period of the expansion.

In investigating the period of a rational number, we have come back to a concept first defined in Section 4.3, namely, the order of one integer modulo another. Recall that if $(x, a) = 1$, then the order of $x \pmod{a}$ is the least positive integer j such that

$$x^j \equiv 1 \pmod{a}.$$

In the present case, we have found that j, which is the period of the decimal expansion of b/a, is also the order of 10 modulo a.

A pertinent result is Theorem 4.8, which says that the order of an integer x modulo a must divide $\phi(a)$. Applying this theorem with $x = 10$ gives the last line of the following theorem.

Theorem 6.18. *Let a and b be relatively prime integers with $0 \leq b < a$ and $(a, 10) = 1$. Suppose the decimal expansion of b/a has period j. Then j is the order of 10 modulo a. In particular, j divides $\phi(a)$.*

Example. Find the period of the decimal expansion of $2/7$.

Although we could just divide 7 into 2, we will instead compute the order of $10 \pmod{7}$. This must be a divisor of $\phi(7) = 6$. We have

$$10^1 \equiv 3, \quad 10^2 \equiv 9 \equiv 2, \quad \text{and} \quad 10^3 \equiv 3 \cdot 2 \equiv 6,$$

where all the congruences are modulo 7. Since the order of $10 \pmod{7}$ is not 1, 2, or 3, it must be 6. In fact $2/7 = .\overline{285714}$. ◇

Notice that the period of the expansion of b/a depends only on the denominator a (assuming $(a, b) = 1$, of course).

Decimals for $1/n$ with Maximal Period

People who play around with numbers have often considered the question of when the period of the decimal representation of $1/n$ has length $n-1$, as, for example, $1/7 = .\overline{142857}$, which has period 6. Since we know (assuming that $(n,10) = 1$) the period must be a divisor of $\phi(n)$, which is $n-1$ only when n is prime, this happens only for prime values of n. Having n prime is not sufficient for $1/n$ to have period $n-1$, however. For example, $1/11 = .090909\ldots = .\overline{09}$, which has period 2. This is to be expected, since $10 \equiv -1 \pmod{11}$, so the order of 10 (mod 11) is 2.

In general, if $(g,n) = 1$, then the order of $g \pmod{n}$ is a divisor of $\phi(n)$; the question is when it actually equals $\phi(n)$.

DEFINITION. primitive root

Let n be a positive integer and suppose $(g,n) = 1$. We say that g is a **primitive root** modulo n in case the order of $g \pmod{n}$ is $\phi(n)$.

Examples. We have seen that 10 is a primitive root (mod 7) but not a primitive root (mod 11). The number 11 does have a primitive root, however, namely, $g = 2$. Since the order of 2 (mod 11) must be a divisor of $\phi(11) = 10$, in order to check this it suffices that neither 2^1, $2^2 = 4$, nor $2^5 = 32$ is congruent to 1 (mod 11).

Some integers have no primitive roots. For example if $n = 8$, the only candidates are $g = 1$, 3, 5, and 7. Of course the order of 1 is 1, while the other three numbers have order 2. No integer has order $\phi(8) = 4 \pmod{8}$. ◇

From the above discussion if $(n,10) = 1$, the decimal for $1/n$ has period $n-1$ if and only if n is prime and 10 is a primitive root $\pmod{n}$. Unfortunately the latter characterization is not really helpful, and not much is known about which primes have 10 as a primitive root. It is not even known if there are infinitely many such primes.

Problems for Section 6.4.

A

In the first eight problems, write the given rational number as a periodic decimal.

1. 8/33
2. 13/27
3. 5/14
4. 17/81
5. 6/13
6. 221/37
7. 81/52
8. 2/105

In the next eight problems, write the given periodic decimal in the form b/a with $(a,b) = 1$.

9. $.\overline{7}$
10. $.0\overline{5}$
11. $.7\overline{3}$
12. $.\overline{171}$
13. $1.3\overline{5}$
14. $2.2\overline{954}$
15. $.\overline{4356}$
16. $2.\overline{079002}$

In the next six problems, determine the period of the decimal representation of the given number.

17. 93/7 18. 100/13 19. 112/21
20. 17/33 21. 3/101 22. 13/77

In the next six problems, for each given prime p determine whether or not the decimal for $1/p$ has period $p-1$.

23. 13 24. 17 25. 19 26. 23 27. 29 28. 31

In the next six problems, determine the smallest positive integer g, if any, such that g is a primitive root modulo the given integer.

29. 10 30. 12 31. 14 32. 15 33. 16 34. 18

B

35. Find all positive integers a, $(a, 10) = 1$, such that the decimal expansion of $1/a$ has period 2.
36. Find all positive integers a, $(a, 10) = 1$, such that the decimal expansion of $1/a$ has period 3
37. Show that if b, r, and s are nonnegative integers, then $b/2^r5^s$ has a terminating decimal expansion.
38. Show that if b/a has a terminating decimal expansion, where a and b are positive integers and $(a, b) = 1$, then $a = 2^r5^s$ for nonnegative integers r and s.
39. Show that if $(a, b) = (a, b') = 1$, with a, b, and b' positive integers, if 10 is a primitive root modulo a, and if

$$b/a = e_1 \cdots e_m.d_1 \cdots d_k\overline{d_{k+1} \ldots d_{k+j}},$$

then there exist digits $f_1, \ldots, f_r, g_1, \ldots, g_s$ such that

$$b'/a = f_1 \cdots f_r.g_1 \cdots g_s\overline{d_{k+1} \ldots d_{k+j}}.$$

40. Let a and b be positive integers with $(a, b) = 1$. Suppose $a = 2^r5^sa'$, where $(a', 10) = 1$. Show that the period of the decimal expansion of b/a is the order of $10 \pmod{a'}$.
41. Suppose $(10, m) = 1$, $m > 0$. Show that there exists an integer k such that the decimal representation of k consists only of 9s, and such that $m \mid k$. (**Hint:** Consider $10^j \equiv 1 \pmod{m}$.)
42. Suppose $(10, m) = 1$, $m > 0$. Show that there exists an integer k such that the decimal representation of k consists only of 1s, and such that $m \mid k$.

C

The next three problems concern b-ary expansions, that is, expansions of the form

$$.a_1a_2 \cdots = \frac{a_1}{b} + \frac{a_2}{b^2} + \cdots ,$$

where $b > 1$ and $0 \le a_i < b$ for all i.

43. Show that a b-ary expansion of $1/n$ terminates if and only if n divides some integral power of b.
44. Show that if $(n, b) = 1$, then the period of a b-ary expansion of $1/n$ equals the order of $b \pmod{n}$.
45. Show that if a positive real number has more than one b-ary expansion, then it has exactly two, and one of these is terminating.

6.5 Lagrange's Theorem and Primitive Roots

The main purpose of this section is to prove that if p is prime, then there exists a primitive root modulo p, that is, an element of order $\phi(p) = p - 1 \pmod{p}$. We know that if $(g, p) = 1$, then g has some order $k \pmod{p}$, and that k divides $p-1$. In the course of our proof, we will actually determine how many elements any reduced residue system $\pmod{p}$ has of each possible order.

Let us denote by $N(k)$ the number of elements in a reduced residue system $\pmod{p}$ of order k. For example in Section 4.3, we saw that for $p = 7$ the numbers 1, 2, 3, 4, 5, and 6 have orders 1, 3, 6, 3, 6, and 2, respectively. Thus in this case $N(1) = 1$, $N(2) = 1$, $N(3) = 2$, and $N(6) = 2$. Of course, if k does not divide $p - 1$ then $N(k) = 0$. Since a reduced residue system $\pmod{p}$ has $p - 1$ elements, each of which has some order dividing $p - 1$, we have

$$\sum_{k|p-1} N(k) = p - 1. \tag{6.7}$$

In our example, $N(1) + N(2) + N(3) + N(6) = 1 + 1 + 2 + 2 = 6 = 7 - 1$.

It turns out that $N(k)$ cannot be too large. For example, always $N(1) = 1$, since only the element of a reduced residue system congruent to $1 \pmod{p}$ can have order 1.

How many elements can have order 2? If g has order $2 \pmod{p}$, then $g^2 \equiv 1 \pmod{p}$. This means $p | g^2 - 1 = (g - 1)(g + 1)$. Since p must divide one or the other of these factors, either $g \equiv 1$ or $g \equiv -1 \pmod{p}$. Since 1 has order 1 and -1 has order $2 \pmod{p}$ for $p > 2$, we have $N(2) = 1$. (If $p = 2$ then $-1 \equiv 1$ and we have $N(2) = 0$.) Note that the argument depends on the assumption that p is prime. The integers 3, 5, and 7 all have order $2 \pmod{8}$.

In the same way, if g has order $3 \pmod{p}$, then $g^3 \equiv 1 \pmod{p}$, so p divides $g^3 - 1 = (g - 1)(g^2 + g + 1)$. Thus if g is not congruent to $1 \pmod{p}$, it must satisfy the quadratic congruence

$$x^2 + x + 1 \equiv 0 \pmod{p}.$$

By Theorem 5.3 this congruence has at most two solutions. We see that $N(3)$ is at most 2.

Lagrange's Theorem

In general, if g has order $k \pmod{p}$, then g satisfies

$$x^k - 1 \equiv 0 \pmod{p}.$$

We would like to be able to say that this congruence, involving a polynomial of degree k, has at most k solutions in any complete residue system $\pmod{p}$, just as from algebra, we know that a polynomial equation of degree k has at most k solutions. First, it is necessary to build up a small amount of theory about polynomials.

Lemma 6.19. *Let f be a polynomial with integer coefficients, and let r be any integer. Then there exists a polynomial g with integer coefficients such that*

$$f(x) - f(r) = (x - r)g(x).$$

Proof. Let $f(x) = a_n x^n + a_{n-1}x^{n-1} + \cdots + a_0$. Then

$$\begin{aligned} f(x) - f(r) &= a_n(x^n - r^n) + a_{n-1}(x^{n-1} - r^{n-1}) + \cdots + a_1(x - r) \\ &= (x - r)g(x), \end{aligned}$$

since it is easily seen that for each positive integer j

$$x^j - r^j = (x - r)(x^{j-1} + x^{j-2}r + x^{j-3}r^2 + \cdots + r^{j-1})$$

by multiplying out the right side and cancelling. □

Lemma 6.20. *If f is a nonconstant polynomial with integer coefficients, m is a positive integer, and r is an integer such that $f(r) \equiv 0 \pmod{m}$, then there exists a polynomial g with integer coefficients with the same leading coefficient as f such that the corresponding coefficients of $f(x)$ and $(x - r)g(x)$ are all congruent $\pmod{m}$ and such that the degree of g is one less than that of f.*

Proof. From Lemma 6.19, there exists an integral polynomial g such that

$$f(x) - f(r) = (x - r)g(x).$$

Clearly g has the same leading coefficient as f and degree one less. The coefficients of $(x-r)g(x)$ are the same as those of $f(x)$ except for the constant terms, which differ by $f(r) \equiv 0 \pmod{m}$. □

If two integral polynomials P_1 and P_2 have corresponding coefficients congruent $\pmod{m}$, we will write

$$P_1(x) \sim P_2(x) \pmod{m}.$$

For example in Lemma 6.20 we have $f(x) \sim (x-r)g(x) \pmod{m}$. Notice that if $P_1(x) \sim P_2(x) \pmod{p}$, then for any integer r we have $P_1(r) \equiv P_2(r) \pmod{p}$.

Theorem 6.21 (Lagrange's theorem). *Let f be a polynomial of degree n with integer coefficients, and let p be a prime. Suppose p does not divide the leading coefficient of f. Then the congruence*

$$f(x) \equiv 0 \pmod{p} \tag{6.8}$$

has at most n solutions in any complete residue system modulo p.

Proof. Suppose $r_1, \ldots, r_{n+1}$ are incongruent solutions of (6.8). By Lemma 6.20 we can write

$$f(x) \sim (x - r_1)g(x) \pmod{p},$$

where g has the same leading coefficient as f but degree 1 less. Then for $i = 2, 3, \ldots, n+1$,

$$f(r_i) \equiv (r_i - r_1)g(r_i) \equiv 0 \pmod{p}.$$

Since r_i is not congruent to $r_1 \pmod{p}$ for $i > 1$, we see that $r_2, r_3, \ldots, r_{n+1}$ all satisfy

$$g(x) \equiv 0 \pmod{p}.$$

Thus in the same way,

$$g(x) \sim (x - r_2)h(x) \pmod{p},$$

where h has the same leading coefficient as g and degree one less. This means

$$f(x) \sim (x - r_1)(x - r_2)h(x) \pmod{p}.$$

If we continue in this way, we get

$$f(x) \sim (x - r_1)(x - r_2) \cdots (x - r_n)c \pmod{p}, \tag{6.9}$$

where c is the leading coefficient of f. But then $f(r_{n+1}) \equiv 0 \pmod{p}$ is impossible, since when r_{n+1} is substituted into (6.9) for x none of the factors on the right is divisible by p. □

The Number of Elements of a Given Order Modulo p

In light of Lagrange's theorem, we see that since any element of order $k \pmod{p}$ satisfies

$$x^k \equiv 1 \pmod{p}, \tag{6.10}$$

there can be at most k such elements in any reduced residue system $(\text{mod } p)$. Actually we can do a little better than this. Suppose $N(k)$ is not 0, so there exists an integer a with order $k \pmod{p}$. From Theorem 4.6, we know that the elements $a, a_2, \ldots, a^k$ are all distinct $(\text{mod } p)$. It is easy to see that each of them also satisfies (6.10), since

$$(a^j)^k = (a^k)^j \equiv 1^j \equiv 1 \pmod{p}.$$

Since Lagrange's theorem says that 6.10 can have at most k distinct solutions $(\text{mod } p)$, the numbers $a, a^2, \ldots, a^k$ must constitute all of them.

Not all of these powers of a have order k, however. We claim that a^j has order $k \pmod{p}$ if and only if $(j, k) = 1$. To see this first suppose that $(j, k) = d > 1$. Then

$$(a^j)^{k/d} = (a^k)^{j/d} \equiv 1^{j/d} \equiv 1 \pmod{p},$$

from which we see that the order of a^j does not exceed $k/d < k$.

If $(j, k) = 1$, on the other hand, we will show that for no positive integer $h < k$ do we have

$$(a^j)^h \equiv 1 \pmod{p}. \tag{6.11}$$

For if we suppose that 6.11 holds, then by Theorem 4.6 we have $k|jh$. But since $(j, k) = 1$ this means that $k|h$. Thus $h \geq k$.

Since we have proved that if a has order k then the elements of order k in some reduced residue system (mod p) are exactly the integers a^j, $1 \leq j \leq k$, with $(j, k) = 1$, we have established the following result.

Lemma 6.22. *Suppose p is prime and k divides $p - 1$. Then $N(k)$ is either* 0 *or $\phi(k)$.*

Example. The reader should check that the following table correctly gives the order of each of the integers 1, 2, ... , 12 (mod 13).

element	1	2	3	4	5	6	7	8	9	10	11	12
order	1	12	3	6	4	12	12	4	3	6	12	2

The computation of orders can be simplified by what we learned in the proof of the lemma. For example, once we know that the order of 2 (mod 13) is 12, then the elements $6 \equiv 2^5$, $11 \equiv 2^7$, and $7 \equiv 2^{11}$ must also have order 12 (and so be primitive roots modulo 13), since 5, 7, and 11 are relatively prime to 12.

In this case, we have $N(1) = 1$, $N(2) = 1$, $N(3) = 2$, $N(4) = 2$, $N(6) = 2$, and $N(12) = 4$. Notice that for each divisor k of $\phi(13) = 12$ we have $N(k) = \phi(k)$. We will prove that this is always the case. ♢

From Lemma 6.22 and formula (6.7), we can now say that

$$p - 1 = \sum N(k) \leq \sum \phi(k),$$

where both summations run over the positive divisors k of $p - 1$. By Theorem 3.11 the summation on the right is exactly $p - 1$, and so we have

$$p - 1 = \sum_{k|p-1} N(k) \leq \sum_{k|p-1} \phi(k) = p - 1.$$

Since we have $p - 1$ on both extremes, all the inequalities must be equalities; that is, $N(k) = \phi(k)$ whenever k divides $p - 1$.

Theorem 6.23. *Suppose p is prime and k divides $p - 1$, $k > 0$. Then any reduced residue system modulo p contains exactly $\phi(k)$ elements having order k* (mod p)*. In particular, there exist primitive roots modulo p.*

Notice that this is merely an existence theorem, and does not give any way to find a primitive root for a particular prime. It has not even been proved that there is any fixed number, such as 2, which is a primitive root for infinitely many primes, although this seems likely.

Moduli Having Primitive Roots

It can be shown that the modulus $n > 1$ has a primitive root exactly in the following cases:

1. $n = p^r$ for some odd prime p.
2. $n = 2p^r$ for some odd prime p.
3. $n = 2$ or $n = 4$.

For example, among the integers from 20 to 30, precisely 23, 25, 27, 29 (all case (1)), 22, and 26 (both case (2)) have primitive roots. This statement will not be proved here, although some cases will be found in the exercises for this section.

Indices

Suppose g is a primitive root modulo n. Then the powers of g run through a reduced residue system (mod n). This means that elements of this reduced residue system can be multiplied by adding the corresponding exponents of g.

As an example, we take $n = 13$ and $g = 2$. The following table gives the least residue of each number 2^j, $j = 0, 1, \ldots, 11$.

j	0	1	2	3	4	5	6	7	8	9	10	11
$2^j \pmod{13}$	1	2	4	8	3	6	12	11	9	5	10	7

Definition. index

If g is a primitive root (mod n) and $(a, n) = 1$, by the **index** of a we mean the number j, $0 \le j < \phi(n)$, such that $g^j \equiv a \pmod{n}$. (Of course, the index of a depends on n and g also.) For example, from the above table we see that the index of 8 is 3, and the index of 3 is 4.

It is useful to turn our table around so as to display the index of each number.

a	1	2	3	4	5	6	7	8	9	10	11	12
index of a	0	1	4	2	9	5	11	3	8	10	7	6

This table of indices can be used to multiply (mod 13) as follows. Suppose we want the least residue of $8 \cdot 11$ (mod 13). We have that

$$8 \cdot 11 \equiv 2^3 \cdot 2^7 \equiv 2^{10} \equiv 10 \pmod{13}.$$

In the same way, we compute that

$$5 \cdot 9 \cdot 7 \equiv 2^9 2^8 2^{11} \equiv 2^{28} \equiv (2^{12})^2 2^4 \equiv 1 \cdot 3 \equiv 3 \pmod{13},$$

where at the end we used the fact that $2^{12} \equiv 1 \pmod{13}$ by Euler's theorem. This means that exponents may be reduced modulo 12 (or, in general, modulo $\phi(n)$).

This system replaces multiplication by addition for reduced residue systems (mod n) exactly as logarithms do for real numbers. Likewise, division (mod n) may be replaced by subtraction; but we leave the details for the reader to work out.

Problems for Section 6.5.

A

In the first eight problems, for each given number n tell how many primitive roots there are in a reduced residue system modulo n.

1. 17
2. 19
3. 10
4. 14
5. 101
6. 103
7. 105
8. 107

9. Which of the numbers 50, 51, ... , 59 have primitive roots?
10. Which of the numbers 80, 81, ... , 89 have primitive roots?

In the next four problems, use the tables for $n = 13$ at the end of this section to find the least residue modulo 13 for the given number.

11. $9 \cdot 12$
12. $11 \cdot 7$
13. $6 \cdot 7 \cdot 8$
14. $5 \cdot 7 \cdot 9$

15. Make a table like the last one in this section for the modulus 7 and primitive root 3.
16. Make a table like the last one in this section for the modulus 11 and primitive root 6.

B

True-False. In the next nine problems, tell which statements are true, and give counterexamples for those that are false. Assume that a and b are primitive roots modulo m.

17. The integer ab is a primitive root (mod m).
18. The integer $-a$ is a primitive root (mod m).
19. The integer $a+b$ is a primitive root (mod m).
20. The integer $a+b$ is not a primitive root (mod m).
21. The integer a^2 is not a primitive root (mod m).
22. If $b \equiv a^k \pmod{m}$, then $(m, k) = 1$.
23. If $m > 2$, then $a^{\phi(m)/2} \equiv -1 \pmod{m}$.
24. If c is a solution to $ac \equiv 1 \pmod{m}$, then c is a primitive root (mod m).
25. If $c^{\phi(m)/2} \equiv -1 \pmod{m}$, then c is a primitive root (mod m).

26. Let p be an odd prime and g a primitive root (mod p). Prove g is not a quadratic residue (mod p).
27. Show that if g is a primitive root (mod p), p prime, and $(a, p) = 1$, then there exists a unique integer t, $0 \leq t < p-1$, such that $g^t \equiv a \pmod{p}$.
28. Let g, a, p, and t be as in the previous problem. Show that a is a quadratic residue (mod p) if and only if t is even.
29. Show that Lagrange's theorem does not hold if the modulus is not restricted to be a prime.

30. Show that Lagrange's theorem does not hold if the restriction that the leading coefficient of f is not divisible by p is dropped.
31. Show that if the Fermat number F_r is prime, then any quadratic nonresidue (mod F_r) is a primitive root.
32. Let a and b be positive integers with $(a, b) = 1$. Show that if g is a primitive root (mod ab), then g is a primitive root modulo both a and b.
33. Show that the only positive powers of 2 having primitive roots are 2 and 4.
34. C Let p and $q = 2p + 1$ be primes, with $p \equiv 3 \pmod 4$. Show that every quadratic nonresidue (mod q) not congruent to $-1 \pmod q$ is a primitive root (mod q).
35. With the hypotheses of the previous problem, show that -2 is a primitive root for q.
36. Suppose the Fermat number F_r is prime, $r > 0$. Show that 3 is a primitive root (mod F_r).
37. Let g be a primitive root (mod n), n odd. Let h be g or $g + n$, whichever is odd. Show that h is a primitive root (mod $2n$).
38. Let $p \equiv 1 \pmod 4$ be prime. Show that if g is a primitive root (mod p), then so is $-g$.
39. Let $(a, b) = 1$ and suppose g has order r (mod a) and s (mod b). Show that the order of g (mod ab) is $[r, s]$.
40. Show that if ab has a primitive root, with $(a, b) = 1$, then $a < 3$ or $b < 3$. (**Hint:** Use the last problem to show that $\phi(a)$ and $\phi(b)$ are relatively prime.)
41. Show that if n has a primitive root and if the odd prime p divides n, then n is of the form p^j or $2p^j$.

Paul Erdös

1913–1996

The only mathematician to rival Euler's productivity has been the Hungarian Paul Erdös (pronounced AIR-dush) with more than 1500 publications and 450 collaborators. Erdös was famous for his problems, mostly in combinatorics and number theory, continually generated and freely shared with others. Most of his life was spent traveling over the world with a small suitcase. He would drop in on mathematical acquaintances unannounced and spend days or weeks prodding the local mathematical community with questions and ideas. His talks often had almost the same title: "Some problems in —," but the content varied as solved problems were replaced by new conjectures, peppered with wry comments about the deterioration of his brain and imminent death.

Erdös produced answers as well as questions, beginning with his proof as a college freshman of "Bertrand's postulate," which says that if $n > 1$, there is always a prime strictly between n and $2n$. (So 3 is between 2 and 4, 5 is between 3 and 6, etc.) The theorem had been proved more than 50 years earlier by the Russian Chebyshev, but Chebyshev's proof was long and involved heavy analysis, while Erdös's was short and could be understood by an undergraduate.

Probably Erdös's best-known accomplishment was also a new proof of an established theorem, namely, the 1948 elementary proof by him and Atle Selberg of the prime number theorem, explained in Section 2.5. Here "elementary" is used in a technical sense, not meaning short or simple, but rather avoiding the use of deep properties of complex numbers. Unfortunately, the young Norwegian Selberg was as secretive and guarded with his results as Erdös was open. After a falling out, they ended up publishing the proof, to which both made essential contributions, in separate papers.

Erdös ignored the usual distractions of the world for mathematics. Sometimes he would sign over his pay to a friend to manage for him. Ronald Graham, an AT&T mathematician and frequent collaborator, learned to forge Erdös's signature to pay his bills. By leaving everything except mathematics to others, Erdös influenced the field far beyond his own considerable contributions.

PAUL ERDÖS

Chapter 7

Diophantine Equations

Although **Diophantine equations** are named for the Greek mathematician Diophantus, who generally sought rational solutions, we will use the term in its more common modern sense, referring to equations for which integral solutions are desired. Almost nothing is known about the personal life of Diophantus, other than that he lived in Alexandria, perhaps about 250 A.D.; but those of his books that survived set a mathematical standard unsurpassed for many centuries. (See the problem at the end of Section 2.3 for more information about him.) The achievements of Diophantus are even more remarkable when one considers that he did not have the advantage of modern algebraic notation, and so had to describe his methods in words or by means of numerical examples.

We have already considered some Diophantine equations, namely, the equation $ax + by = c$ of Section 1.4, and the equation $x^n + y^n = z^n$ of Fermat's last theorem. Actually Diophantus never seems to have written about linear equations such as the former, perhaps because he thought them too simple.

There are few general theorems concerning Diophantine equations, and special proof techniques are often necessary for each equation. We will consider only a few of the simpler and more interesting examples of Diophantine equations.

7.1 Pythagorean Triples

DEFINITION. Pythagorean triple, primitive Pythagorean triple

We call the triple of positive integers (x, y, z) a **Pythagorean triple** in case it satisfies the Diophantine equation

$$x^2 + y^2 = z^2. \tag{7.1}$$

The triple is said to be **primitive** in case no integer greater than 1 divides all of x, y, and z.

Of course, by the Pythagorean theorem in any right triangle whose sides have integral lengths, these lengths form a Pythagorean triple; and by its converse any such triple corresponds to a right triangle. This accounts for the terminology. The best known Pythagorean triple is $(3, 4, 5)$, which is primitive. Other (nonprimitive) triples such as $(6, 8, 10)$, $(9, 12, 15)$, and $(30, 40, 50)$ can easily be formed from this one, since it is easy to see that if (x, y, z) is a Pythagorean triple, then so is (kx, ky, kz) for any positive k.

We see that any primitive Pythagorean triple (x, y, z) generates infinitely many nonprimitive ones (kx, ky, kz), $k = 2, 3, \ldots$. Conversely, any Pythagorean triple (x, y, z) corresponds to a unique primitive one found by dividing x, y, and z by their greatest common divisor. Thus we will concentrate our attention on primitive triples.

Other primitive Pythagorean triples exist, for example, $(12, 5, 13)$, which the reader should check. In fact, there are infinitely many. This can be demonstrated by the following simple geometric argument. Let n be any odd integer greater than 1. Then n^2 is also odd, and so of the form $2m + 1$. Arrange $n^2 = 2m + 1$ unit squares outside two sides of an m-by-m square as follows (in the example we have taken $n = 3$ so $2m + 1 = 9$ and $m = 4$).

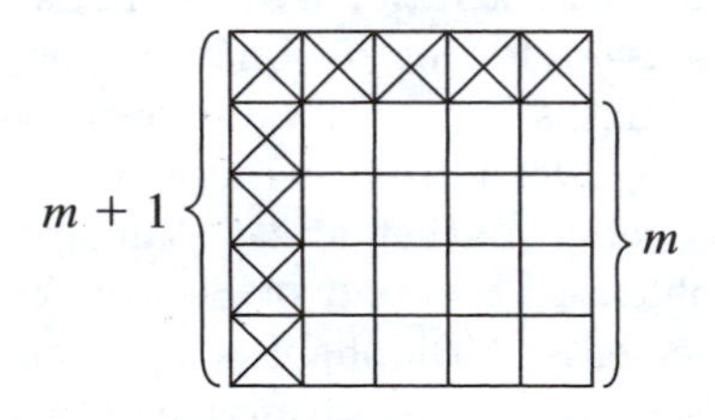

Then

$$\text{(area of inner square)} + \text{(area of border)} = \text{(area of large square)},$$

or

$$m^2 + n^2 = (m+1)^2.$$

This shows that $(m, n, m + 1)$ is a Pythagorean triple, primitive since m and $m + 1$ are relatively prime. Since n was an arbitrary odd integer greater than 1, there exist infinitely many such triples.

Examples. The case illustrated above leads to the triple $(4, 3, 5)$, while taking $n = 5$ makes $2m + 1 = 25$ and $m = 12$. This gives the already-mentioned triple $(12, 5, 13)$. ◊

Finding All Primitive Solutions

One might guess that all primitive Pythagorean triples arise from the above method, but such is not the case. For example, $(8, 15, 17)$ is Pythagorean, since $64 + 225 = 289 = 17^2$. Yet a triple (x, y, z) produced by the geometric method always has z just 1 more than x or y, which is not the case for $(8, 15, 17)$. We will find a slightly more complicated way of generating all Pythagorean triples in this section.

Lemma 7.1. *If (x, y, z) is a primitive Pythagorean triple, then x, y, and z are relatively prime in pairs, one of x and y is even and the other odd, and z is odd.*

Proof. Suppose $(x, y) = d$. If $d > 1$, then some prime p divides d. From equation (7.1) and Theorem 1.2, we see that p divides z^2. But then by the fundamental theorem of arithmetic p divides z also, contradicting the assumption that (x, y, z) is primitive. Similar proofs show that $(x, z) = (y, z) = 1$ also.

Now by the first part of this lemma x and y cannot both be even. We claim that they cannot both be odd either. For if they were, then (7.1) would imply that z was even. But if x and y are odd and z is even, we have

$$\begin{aligned} x &\equiv 1 \text{ or } 3 \pmod 4, \\ y &\equiv 1 \text{ or } 3 \pmod 4, \\ z &\equiv 0 \text{ or } 2 \pmod 4. \end{aligned}$$

Squaring these congruences gives

$$\begin{aligned} x^2 &\equiv 1 \pmod 4, \\ y^2 &\equiv 1 \pmod 4, \\ z^2 &\equiv 0 \pmod 4. \end{aligned}$$

But adding the first two congruences gives

$$x^2 + y^2 \equiv 2 \pmod 4,$$

which contradicts (7.1) and the third congruence. Thus one of x and y is even and the other odd, and z must be odd. □

Now we will turn to a general analysis of primitive Pythagorean triples. We have seen from the lemma that one of x and y is even and the other odd. Since (7.1) is symmetric in x and y we may label a solution so that x is even, say $x = 2t$, and y is odd. Substituting $x = 2t$ in (7.1) gives

$$4t^2 + y^2 = z^2,$$

or

$$4t^2 = z^2 - y^2 = (z + y)(z - y).$$

Since y and z are both odd the two factors on the right are even, so we can write

$$t^2 = \frac{z+y}{2} \cdot \frac{z-y}{2}, \tag{7.2}$$

where the factors on the right are not only integers, but relatively prime, since any common factor greater than 1 would have to divide both their sum, which is z, and their difference, which is y, contradicting the lemma.

Now since each prime dividing the left side of the last equation does so to an even power, the same goes for the right side of the equation. Because no prime can divide both of the right-hand factors, each contains only even prime powers and so is a perfect square, say

$$\frac{z+y}{2}=u^2 \quad \text{and} \quad \frac{z-y}{2}=v^2,$$

where we may assume u and v are positive.

Adding these two equations gives $z=u^2+v^2$, and subtracting them gives $y=u^2-v^2$, from which we see that $u>v$ since $y>0$. Note that since z is odd u and v cannot be both even or both odd. Since we saw above that u^2 and v^2 were relatively prime, the same goes for u and v.

Finally, from (7.2)

$$t^2=u^2v^2,$$

and so $x=2t=2uv$.

Theorem 7.2 (the Pythagorean triple theorem). *Suppose that (x,y,z) is a primitive Pythagorean triple with x even. Then there exist relatively prime positive integers u and v, not both odd, with $u>v$, such that*

$$x=2uv, \quad y=u^2-v^2, \quad \textit{and} \quad z=u^2+v^2.$$

Conversely, if u and v are any pair of integers as described in the previous sentence, and if x, y, and z are defined by these three equations, then (x,y,z) is a primitive Pythagorean triple with x even.

Proof. The first part of the theorem has just been proved. To prove the second part, assume that u and v are as described, and let x, y, and z be defined by the last three equations. Then

$$\begin{aligned} x^2+y^2 &= (2uv)^2+(u^2-v^2)^2 \\ &= 4u^2v^2+u^4-2u^2v^2+v^4 \\ &= u^4+2u^2v^2+v^4=(u^2+v^2)^2=z^2, \end{aligned}$$

and so (7.1) is satisfied.

Clearly x is even. It only remains to show that (x,y,z) is primitive. Suppose some prime p divides both y and z. Then it divides their sum, which is $2u^2$, and their difference, which is $2v^2$. Since by hypothesis one of u and v is even and the other odd, y is odd and so p is not 2. Then p divides both u and v, contradicting the assumption that u and v are relatively prime. □

Examples. The following table gives Pythagorean triples corresponding to some small values of u and v.

u	v	x	y	z
2	1	4	3	5
3	2	12	5	13
4	1	8	15	17
4	3	24	7	25

Note that any pair of positive integers u and v with $u > v$ will generate a Pythagorean triple (x, y, z) from the equations of Theorem 7.2; the extra conditions on u and v are necessary to guarantee that the triple is primitive. ◇

Example A. Find all primitive Pythagorean triples with $x = 28$.

Since $x = 2uv$, we have $uv = 14$. The only possibilities for u and v satisfying the theorem are $u = 14$, $v = 1$, and $u = 7$, $v = 2$. These lead to the triples $(28, 195, 197)$ and $(28, 45, 53)$, respectively. ◇

Example B. Find all primitive Pythagorean triples with $y = 15$.

We need to solve the equation $u^2 - v^2 = 15$. Notice that there are only finitely many solutions possible, since the difference between consecutive squares gets larger and larger.

In fact, $15 = (u + v)(u - v)$, and the factors on the right must be either 15 and 1 or else 5 and 3. Solving the equations

$$u + v = 15, \quad u - v = 1$$

gives $u = 8$ and $v = 7$, leading to the triple $(112, 15, 113)$; while the other case gives $u = 4$ and $v = 1$, leading to $(8, 15, 17)$. ◇

Example C. Find all Pythagorean triples with $x = 28$.

This looks just like Example A until we notice that the condition that the triples be primitive has been dropped. A nonprimitive solution will be a triple $(28, y, z) = (ka, kb, kc)$, where (a, b, c) is a primitive Pythagorean triple and a is some divisor of 28. We cannot even be sure that a is even.

The divisors of 28 are 1, 2, 4, 7, 14, and 28, and we try each of these in turn as elements of a primitive Pythagorean triple. Let us start with the even divisors, which would correspond to $2uv$ in a primitive triple. The equation $2uv = 2$ leads to $uv = 1$, which is impossible since u and v are not both odd. The equation $2uv = 14$ is impossible in a similar way. However, $2uv = 4$ says $uv = 2$, which works with $u = 2$ and $v = 1$. This yields the primitive triple $(4, 3, 5)$. We multiply this by 7 to get the nonprimitive triple $(28, 21, 35)$. The equation $2uv = 28$ was already solved in Example A, yielding the primitive triples $(28, 195, 197)$ and $(28, 45, 53)$.

Now let us consider the odd divisors of 28. To appear as y in a primitive solution they would have to be of the form $u^2 - v^2$. The equation $1 = u^2 - v^2$ is impossible, but $7 = u^2 - v^2 = (u+v)(u-v)$ has the solution $u = 4$, $v = 3$, leading to the primitive triple $(24, 7, 25)$. We multiply this by 4 to get the nonprimitive triple $(96, 28, 100)$. Since it was x that was supposed to be 28 we rearrange this to $(28, 96, 100)$.

We have found four answers to our problem: $(28, 21, 35)$, $(28, 195, 197)$, $(28, 45, 53)$, and $(28, 96, 100)$. ◇

Problems for Section 7.1.

A *The first four problems refer to the geometric method given at the beginning of this section. What Pythagorean triples are generated by the following values of n?*

1. $n = 7$
2. $n = 9$
3. $n = 11$
4. $n = 13$

In the next four problems, find the triples (x, y, z) generated by the given u and v.

5. $u = 6, v = 1$
6. $u = 6, v = 5$
7. $u = 5, v = 2$
8. $u = 5, v = 4$

In the next four problems, find the u and v generating the given triple (x, y, z).

9. $(24, 7, 25)$
10. $(48, 55, 73)$
11. $(36, 77, 85)$
12. $(140, 51, 149)$

In the next seven problems, assume (x, y, z) is a primitive Pythagorean triple generated by u and v. Find the three missing variables.

13. $x = 84, y = 13$
14. $u = 7, x = 56$
15. $u = 9, y = 17$
16. $v = 4, z = 97$
17. $u = 8, z = 89$
18. $v = 7, y = 15$
19. $v = 2, x = 36$

B

In the next seven problems, find all primitive Pythagorean triples with the given value of one variable.

20. $x = 12$
21. $x = 24$
22. $y = 9$
23. $y = 21$
24. $y = 20$
25. $z = 125$
26. $z = 65$

In the next four problems, find all Pythagorean triples with the given value of one variable.

27. $x = 16$
28. $x = 40$
29. $y = 35$
30. $z = 85$

31. Show that if (x, y, z) is a primitive Pythagorean triple, then $(x, z) = (y, z) = 1$.
32. Give an example of integers $x, y,$ and z such that no integer greater than 1 divides all three, but none of (x, y), (x, z), and (y, z) is 1.
33. Each Pythagorean triple found by the geometric method given at the beginning of this section corresponds to a triple generated by u and v as in the Pythagorean triple theorem. What are u and v in terms of the odd number n?

C

34. Show that if $x \equiv 2 \pmod 4$, then there is no primitive Pythagorean triple (x, y, z).

35. Suppose a circle is inscribed in a right triangle whose sides have integral lengths. Show that the radius of the circle is an integer.

In the next five problems, assume that x, y, and z are positive integers satisfying

$$x^2 + 2y^2 = z^2, \tag{7.3}$$

and that no integer greater than 1 divides all of x, y, and z. These problems parallel the proof of the Pythagorean triple theorem (with some variations).

36. Show that x, y, and z are relatively prime in pairs.
37. Show that x and z are odd and y is even.
38. Show that the integers $(z + x)/2$ and $(z - x)/2$ are relatively prime, and exactly one of them is even.
39. Define u to be the the square root of half of whichever of the two integers in the previous problem is even, and v to be the square root of whichever of them is odd. Prove that u and v are integers and $(2u, v) = 1$.
40. Prove that $x = |2u^2 - v^2|$, $y = 2uv$, and $z = 2u^2 + v^2$.

41. Show that u and v are any positive integers, and if x, y, and z are defined by the equations of the previous problem, then they satisfy (7.3). Furthermore, if $(2u, v) = 1$, then x, y, and z have no common divisor greater than 1.

7.2 Sums of Two Squares

In the last section, we saw that for each pair of positive integers u, v with $u > v$, defining

$$x = 2uv, \quad y = u^2 - v^2, \quad z = u^2 + v^2$$

produces a Pythagorean triple (x, y, z). In order for this triple to be primitive, u and v must satisfy additional restrictions. It is easy to see from this that each integer $a > 2$ satisfies an equation of the form

$$a^2 + b^2 = c^2.$$

For if a is even, say $a = 2m$, then taking $u = m$ and $v = 1$ produces a triple with $a = x$. If a is odd, on the other hand, the reader should check that taking $u = (a+1)/2$ and $v = (a-1)/2$ produces a Pythagorean triple with $a = y$.

The situation is more complicated if a triple with a specified value of z is to be produced. Needed are integers u and v such that $u^2 + v^2 = z$. That there are infinitely many values of z for which this equation *cannot* be satisfied is a consequence of the following simple but important fact, which can be checked by squaring the elements in any complete residue system (mod 4).

Theorem 7.3. *If k is any integer, then*

$$k^2 \equiv 0 \text{ or } 1 \pmod 4.$$

From this theorem we see that for any integers u and v we have

$$u^2 + v^2 \equiv 0, 1, \text{ or } 2 \pmod 4.$$

Thus no integer congruent to 3 (mod 4) can be written as the sum of two squares.

The Equation $n = u^2 + v^2$

We will investigate the general question of which integers n can be written as the sum of two squares. Being congruent to something other than 3 (mod 4) is not sufficient for an integer to be expressible this way. For example, it is easily checked that

$$u^2 + v^2 = 6$$

has no solution.

For convenience, we will allow u and v to be 0 or even negative. (Of course, if u and v are solutions then so are $-u$ and $-v$.) For n up to 15, the representable numbers are $1 = 0^2 + 1^2$, $2 = 1^2 + 1^2$, $4 = 0^2 + 2^2$, $5 = 1^2 + 2^2$, $8 = 2^2 + 2^2$, $9 = 0^2 + 3^2$, $10 = 1^2 + 3^2$, and $13 = 2^2 + 3^2$.

Of great importance in the question of which integers are the sums of two squares is the identity

$$(a^2 + b^2)(c^2 + d^2) = (ac + bd)^2 + (ad - bc)^2, \tag{7.4}$$

which can be checked by multiplying out both sides.

Lemma 7.4. *If n_1 and n_2 can each be written as the sum of two squares, then so can $n_1 n_2$.*

Proof. Write $n_1 = a^2 + b^2$ and $n_2 = c^2 + d^2$ and apply (7.4). □

Example. Express 130 as the sum of two squares.

We know this can be done since $130 = 10 \cdot 13$, and we saw that $10 = 1^2 + 3^2$ and $13 = 2^2 + 3^2$. In fact, taking $a = 1$, $b = 3$, $c = 2$, and $d = 3$ we see

$$130 = (ac + bd)^2 + (ad - bc)^2 = (2 + 9)^2 + (3 - 6)^2 = 11^2 + (-3)^2.$$

◇

Lemma 7.4 suggests trying to determine which *primes* can be written as the sum of two squares. Up to 15 they are exactly 2, 5, and 13. Further investigation reveals that $17 = 1^2 + 4^2$ is the sum of two squares, but 19 is not; 23 is not, but $29 = 2^2 + 5^2$. The reader is encouraged to form his or her own conjecture before reading on.

The proof given for the next lemma uses the result from Chapter 5 that if p is an odd prime, then $x^2 \equiv -1 \pmod{p}$ is solvable if and only if $p \equiv 1 \pmod{4}$. This is not hard to prove from scratch, however; see problems 27 and 28 of the exercises for this section.

Lemma 7.5. *The prime p can be written as the sum of two squares if and only if $p = 2$ or $p \equiv 1 \pmod{4}$.*

Proof. We know $2 = 1^2 + 1^2$, and we also know that no number congruent to 3 (mod 4), prime or not, can be written as the sum of two squares. Thus it remains merely to show that any prime $p \equiv 1 \pmod{4}$ is the sum of two squares. From part (5) of Theorem 5.7 we know that $(-1/p) = 1$; so there exists an integer x such that $x^2 \equiv -1 \pmod{p}$, or

$$x^2 + 1 \equiv 0 \pmod{p}. \tag{7.5}$$

The argument we now give makes use of the pigeon-hole principle. (See the problems at the end of Section 4.1.) Consider all the integers $rx + s$ with

$$0 \le r < \sqrt{p} \quad \text{and} \quad 0 \le s < \sqrt{p}.$$

Since r and s are allowed to be 0 there are more than $\sqrt{p}$ values of r and more than $\sqrt{p}$ values of s. Thus there are more than $\sqrt{p}\sqrt{p} = p$ pairs r, s. This means at least two of the numbers $rx + s$ must be congruent (mod p), say

$$r_1 x + s_1 \equiv r_2 x + s_2 \pmod{p}.$$

Thus we have

$$(r_1 - r_2)x \equiv s_2 - s_1 \pmod{p}. \tag{7.6}$$

If we let $u = |r_1 - r_2|$ and $v = |s_2 - s_1|$, then from (7.6) we have

$$ux \equiv \pm v \pmod{p}. \tag{7.7}$$

Notice that by the assumption that r_1, s_1 and r_2, s_2 are distinct pairs, u and v cannot both be 0.

Now from (7.5) and (7.7), we have

$$0 \equiv u^2(x^2+1) \equiv (ux)^2 + u^2 \equiv v^2 + u^2 \pmod{p}.$$

This means that $u^2 + v^2$ is some positive multiple of p. But

$$u^2 + v^2 < (\sqrt{p})^2 + (\sqrt{p})^2 = 2p,$$

so we must have $u^2 + v^2 = p$. We have proved that p is the sum of two squares. □

The result that any prime congruent to 1 (mod 4) is the sum of two squares is the key to the analysis of $n = u^2 + v^2$. It was first stated by Albert Girard (1595–1632), and Fermat claimed to have a proof. (He probably did.) The first published proof, however, was by Euler in 1754.

Theorem 7.6. *A positive integer n can be written as the sum of two squares if and only if no prime congruent to 3 modulo 4 appears an odd number of times in the factorization of n into primes.*

Proof. First suppose no prime congruent to 3 (mod 4) appears an odd number of times in the factorization of n. Then we can write

$$n = kp_1p_2\cdots p_t,$$

where all primes congruent to 3 (mod 4) are collected into k, and each prime $p_1, p_2, \ldots, p_t$ is either 2 or congruent to 1 (mod 4). By hypothesis each prime in k appears to an even power, so $k = j^2$. That n is the sum of two squares now follows from the repeated use of Lemma 7.4, since $k = 0^2 + j^2$ and each p_i is a sum of two squares from Lemma 7.5.

To prove the converse assume that $n = u^2 + v^2$. Let $d = (u, v)$, $U = u/d$, $V = v/d$, and $N = n/d^2$. Then U, V, and N are integers, and

$$N = U^2 + V^2. \tag{7.8}$$

Suppose some prime $p \equiv 3 \pmod{4}$ divides n to an odd power. Then p also divides N to an odd power, since dividing n by d^2 removes primes two at a time. If p divided U, then by (7.8) it would also divide V, contradicting the fact that $(U, V) = 1$. Thus there exists an integer x such that $xU \equiv 1 \pmod{p}$. Multiplying (7.8) by x^2 and using the fact that p divides N, we get

$$1 + (xV)^2 \equiv 0 \pmod{p}.$$

Thus $(xV)^2 \equiv -1 \pmod{p}$. But this contradicts Theorem 5.7(5), which says that -1 is not a quadratic residue modulo p for $p \equiv 3 \pmod{4}$. □

Examples. Earlier we found that among the integers from 1 to 15, exactly 1, 2, 4, 5, 8, 9, 10, and 13 could be represented as the sum of two squares. Now we can see why the remaining integers are not so representable. The integers 3, 6, 12, and 15 contain the factor 3 exactly once, 7 and 14 contain 7 exactly once, and 11 contains 11 exactly once.

The integers form 51 to 60 can be analyzed in a similar manner.

$$\begin{aligned}
51 &= 3 \cdot 17 \text{ (note the 3)} \\
52 &= 2^2 13 = 4^2 + 6^2 \\
53 &= 2^2 + 7^2 \\
54 &= 2 \cdot 3^3 \text{ (note the } 3^3\text{)} \\
55 &= 5 \cdot 11 \text{ (note the 11)} \\
56 &= 2^3 7 \text{ (note the 7)} \\
57 &= 3 \cdot 19 \text{ (note the 3 and 19)} \\
58 &= 2 \cdot 29 = 3^2 + 7^2 \\
59 &\equiv 3 \pmod 4 \\
60 &= 2^2 3 \cdot 5 \text{ (note the 3)}
\end{aligned}$$

Problems for Section 7.2.

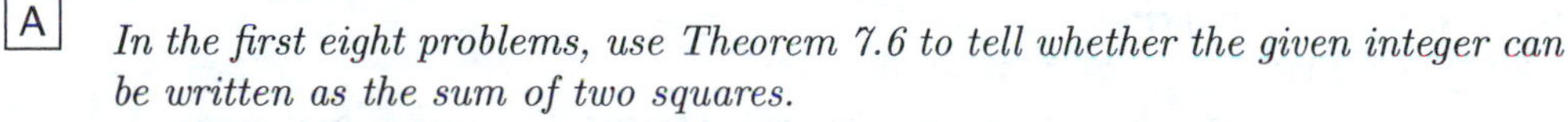

A

In the first eight problems, use Theorem 7.6 to tell whether the given integer can be written as the sum of two squares.

1. 61
2. 62
3. 63
4. 65
5. 68
6. 117
7. $51 \cdot 45$
8. $39 \cdot 54$

In the next six problems, use the identity (7.4) to find positive integers u and v such that $u^2 + v^2$ equals the given integer.

9. $10 \cdot 17$
10. $18 \cdot 5$
11. 85
12. 221
13. 261
14. 637
15. Which integers from 1 to 10 can be written as the sum of three squares?
16. Which integers from 11 to 20 can be written as the sum of three squares?

B

17. Work through the proof of Lemma 7.5 with $p = 5$ and $x = 2$. What are all quadruples r_1, s_1, r_2, s_2 such that $r_1 x + s_1 \equiv r_2 x + s_2 \pmod p$?
18. Repeat the previous problems with $p = 13$ and $x = 5$.
19. Prove that $a^2 \equiv 0, 1,$ or $4 \pmod 8$ for all integers a.
20. Prove that if $n \equiv 7 \pmod 8$ then n cannot be expressed as the sum of three squares.
21. Show that if $4 \mid x^2 + y^2 + z^2$, then x, y, and z are all even.

22. Show that no integer of the form $4^k(8m+7)$ can be written as the sum of three squares. (**Hint:** Use the two previous problems.)
23. Show that if p is a prime congruent to 1 (mod 4), then there exist u and v such that $p = u^2 + v^2$ with $0 < u < v$ and $(u, v) = 1$.
24. Show that if p and q are distinct primes congruent to 1 (mod 4), then $pq = u^2 + v^2$, $0 < u < v$, has two distinct solutions. (**Hint:** Use the previous problem and the identity (7.4).)
25. Show that if $n \equiv 2 \pmod 4$, then n cannot be written as the difference of two squares.
26. Show that any positive integer $n \not\equiv 2 \pmod 4$ can be written as the difference of two squares. (**Hint:** See the discussion at the beginning of this section for the case when n is odd.)

In the next two problems, assume that p is an odd prime. They show that $x^2 \equiv -1 \pmod p$ is solvable if and only if $p \equiv 1 \pmod 4$. Do not use any result from Chapter 5 in proving them.

27. Suppose that x is an integer such that $x^2 \equiv -1 \pmod p$. Show that $(x^2)^{(p-1)/2} \equiv 1 \pmod p$, and conclude that $p \equiv 1 \pmod 4$.
28. Suppose $x^2 \equiv -1 \pmod p$ has no solutions. Let $S = \{1, 2, \ldots, p-1\}$ and show that if $t \in S$ then there exists a unique $t' \in S$, $t' \neq t$, such that $tt' \equiv -1 \pmod p$. Conclude that $(p-1)! \equiv (-1)^{(p-1)/2} \pmod p$ and $p \equiv 3 \pmod 4$.

C

In the next four problems, assume that $p = a^2 + b^2 = c^2 + d^2$, where $0 < a < b$, $0 < c < d$, and p is prime.

29. Show that $a^2d^2 \equiv b^2c^2 \pmod p$.
30. Show that $ad - bc = 0$ or $ad + bc = p$. (**Hint:** Each of a, b, c, and d is $< \sqrt{p}$.)
31. Show that $ad = bc$. (**Hint:** Use (7.4).)
32. Show that $a = c$ and $b = d$, so that the representation of a prime as the sum of two squares is essentially unique. (**Hint:** Show that $(a, b) = (c, d) = 1$.)

7.3 Sums of Four Squares

In the last section, we saw that not every integer can be written as the sum of two squares. Even three squares does not suffice to represent some numbers, for example, 7. In the problems at the end of Section 7.2, it is shown that no integer of the form $4^k(8m+7)$ can be written as the sum of three squares. Actually every positive integer not of this form can be expressed as the sum of three squares, although we will not prove that result. One reason that the analysis of sums of three squares is difficult is that there is no identity similar to (7.4). In fact, it is not true that the product of numbers that are the sum of three squares is also of that form, as shown by the example

$$2 \cdot 14 = (0^2 + 1^2 + 1^2)(1^1 + 2^2 + 3^2) = 28 = 4^1(8 \cdot 0 + 7).$$

It turns out that if four squares are allowed, then every positive integer can be represented. This result, which we will prove in this section, may have been known to Diophantus, although he does not state it explicitly in any of his known writings. Fermat claimed to have a proof of it, but, as was his custom, he published none. Euler worked on the problem for many years, but never found a proof. The first published proof was by Lagrange in 1772, using many of Euler's partial results. Among these was an identity showing that the product of two numbers, each the sum of four squares, is also of that form. We state that identity now; the reader should prove it by multiplying out both sides.

Lemma 7.7. $(a^2+b^2+c^2+d^2)(A^2+B^2+C^2+D^2) = (aA+bB+cC+dD)^2 + (aB-bA+cD-dC)^2 + (aC-bD-cA+dB)^2 + (aD+bC-cB-dA)^2.$

Because of Lemma 7.7, it suffices to prove that each prime can be written as the sum of four squares. But Theorem 7.6 says that all primes except those congruent to 3 (mod 4) can be written as the sum of two squares (and so four squares), so we need only worry about primes of the form $4m+3$.

Lemma 7.8. *Let p be a prime congruent to* 3 (mod 4). *Then there exists an integer k with $0 < k < p$ such that kp is the sum of four squares.*

Proof. Let t be the smallest positive integer that is not a quadratic residue (mod p). Clearly $t > 1$. Then by parts (2) and (5) of Theorem 5.7 we have

$$\left(\frac{-t}{p}\right) = \left(\frac{-1}{p}\right)\left(\frac{t}{p}\right) = (-1)(-1) = 1.$$

This means that $-t$ is a quadratic residue (mod p). Thus there exists an integer x with $x^2 \equiv -t \pmod{p}$, with $0 < x < p/2$. (See Theorem 5.4.)

Now by the definition of t the positive integer $t-1$ must be a quadratic residue (mod p), and so in the same way there exists y such that $y^2 \equiv t-1 \pmod{p}$ and $0 < y < p/2$. Then

$$x^2 + y^2 + 1 \equiv -t + (t-1) + 1 \equiv 0 \pmod{p},$$

and so $x^2 + y^2 + 1 = kp$ for some positive integer k. But then

$$kp = x^2 + y^2 + 1 < \left(\frac{p}{2}\right)^2 + \left(\frac{p}{2}\right)^2 + 1 = \frac{p^2}{2} + 1 < \frac{p^2}{2} + \frac{p^2}{2} = p^2,$$

and so $k < p$. Since $kp = x^2 + y^2 + 1^2 + 0^2$, the lemma is proved. □

Example. We will illustrate this lemma with $p = 7$. The quadratic residues are $1^2 = 1$, $2^2 = 4$, and $3^2 \equiv 2 \pmod{7}$. Thus the smallest positive nonresidue is $t = 3$. Since $-t = -3 \equiv 4 \equiv 2^2 \pmod{7}$, we take $x = 2$. Likewise, $t-1 = 2 \equiv 3^2 \pmod{7}$, so $y = 3$. Then $x^2 + y^2 + 1 = 4 + 9 + 1 = 14 = 2 \cdot 7$, so $k = 2 < p$. ◇

Theorem 7.9 (Lagrange). *Every positive integer can be written as the sum of four squares.*

Proof. We have seen that it suffices to prove that any prime congruent to 3 (mod 4) is the sum of four squares. Let p be such a prime. From Lemma 7.8, there exists a positive integer $k < p$ such that kp is the sum of four squares. (Actually the proof from this point on will work for any odd prime p for which such a k exists.) Of all such integers k, let m be the smallest, so we have

$$mp = a^2 + b^2 + c^2 + d^2, \quad 0 < m < p. \tag{7.9}$$

First, we show m is odd. For if m is even, then of the integers a, b, c, and d exactly none, two, or four must be even. If two of them are even, relabel them so that they are a and b. Then in any case $a - b$, $a + b$, $c - d$, and $c + d$ are all even. But then

$$\left(\frac{a-b}{2}\right)^2 + \left(\frac{a+b}{2}\right)^2 + \left(\frac{c-d}{2}\right)^2 + \left(\frac{c+d}{2}\right)^2 = \frac{a^2 + b^2 + c^2 + d^2}{2} = \frac{m}{2} \cdot p,$$

which contradicts the minimality of m.

Now if $m = 1$, we are done, for then p is the sum of four squares. Assume $m > 1$. Since the m consecutive integers

$$-\frac{m-1}{2}, -\frac{m-1}{2} + 1, \ldots, 0, 1, \ldots, \frac{m-1}{2}$$

form a complete residue system (mod m), we can choose elements A, B, C, and D from among them (so that each is less than $m/2$ in absolute value) such that

$$A \equiv a, \quad B \equiv b, \quad C \equiv c, \quad \text{and} \quad D \equiv d \pmod{m}.$$

Then

$$A^2 + B^2 + C^2 + D^2 \equiv a^2 + b^2 + c^2 + d^2 = mp \equiv 0 \pmod{m}.$$

Let

$$A^2 + B^2 + C^2 + D^2 = rm. \tag{7.10}$$

Note that $r > 0$, for if A, B, C, and D are all 0, then m divides each of a, b, c, and d. But then m^2 divides $a^2 + b^2 + c^2 + d^2 = mp$, which is impossible since $1 < m < p$. Notice also that

$$rm = A^2 + B^2 + C^2 + D^2 < 4\left(\frac{m}{2}\right)^2 = m^2,$$

and so $r < m$.

Now we will make use of the actual identity of Lemma 7.7. From (7.9) and (7.10), we have

$$(mp)(rm) = (a^2 + b^2 + c^2 + d^2)(A^2 + B^2 + C^2 + D^2) = w^2 + x^2 + y^2 + z^2,$$

where

$$\begin{aligned}
w &= aA + bB + cC + dD \equiv a^2 + b^2 + c^2 + d^2 = mp \equiv 0 \pmod{m},\\
x &= aB - bA + cD - dC \equiv ab - ba + cd - dc \equiv 0 \pmod{m},\\
y &= aC - bD - cA + dB \equiv ac - bd - ca + db \equiv 0 \pmod{m},\\
z &= aD + bC - cB - dA \equiv ad + bc - cb - da \equiv 0 \pmod{m}.
\end{aligned}$$

But then

$$rp = \left(\frac{w}{m}\right)^2 + \left(\frac{x}{m}\right)^2 + \left(\frac{y}{m}\right)^2 + \left(\frac{z}{m}\right)^2,$$

which contradicts the minimality of m, since $0 < r < m$. □

Although it is natural to use Theorem 7.6 to reduce the problem to primes congruent to 3 (mod 4), the four-square theorem can also be proved without any knowledge of sums of two squares. All we need to do is include in Lemma 7.8 primes congruent to 1 (mod 4) also. It is also possible to prove Lemma 7.8 (for all odd primes) without using any results on quadratic residues. These proofs are worked out in the exercises.

Problems for Section 7.3.

A

In the first six problems, find all quadruples a, b, c, d with $0 \le a \le b \le c \le d$ and $n = a^2 + b^2 + c^2 + d^2$.

1. $n = 20$
2. $n = 27$
3. $n = 30$
4. $n = 31$
5. $n = 32$
6. $n = 36$

In the next four problems, work through the proof of Lemma 7.8 for the given value of p. What x, y, and k result?

7. $p = 11$ 8. $p = 19$ 9. $p = 23$ 10. $p = 31$

In the next four problems, use $7 = 1^2 + 1^2 + 1^2 + 2^2$, $11 = 0^2 + 1^2 + 1^2 + 3^2$, $10 = 1^2 + 1^2 + 2^2 + 2^2$, *and Lemma 7.7 to express each integer as the sum of four squares.*

11. $7 \cdot 11$ 12. $11 \cdot 10$ 13. $7 \cdot 10$ 14. $77 \cdot 10$

In each of the next four problems, we have mp *written as the sum of four squares, where* $1 < m < p$. *Use the methods of the proof of Theorem 7.9 to write a smaller multiple of* p *as a sum of four squares. (The method will depend on whether* m *is even or odd.)*

15. $p = 19$, $6p = 8^2 + 7^2 + 1^2 + 0^2$ 16. $p = 29$, $4p = 9^2 + 5^2 + 3^2 + 1^2$
17. $p = 11$, $3p = 5^2 + 2^2 + 2^2 + 0^2$ 18. $p = 17$, $5p = 8^2 + 4^2 + 2^2 + 1^2$

B

19. In Lemma 7.8, it is proved that there exist nonnegative integers x and y, each less than $p/2$, such that $x^2 + y^2 + 1 \equiv 0 \pmod{p}$. Show that this holds also for primes $p \equiv 1 \pmod{4}$ without using Theorem 7.6. (**Hint:** $(-1/p) = 1$.)
20. Which integers from 1 to 20 cannot be expressed as the sum of four *positive* squares?
21. Which integers from 21 to 40 cannot be expressed as the sum of four positive squares?
22. Show that 169 can be written as the sum of one, two, three, and four positive squares.
23. Show that if $n > 169$, then n can be written as the sum of five positive squares. (**Hint:** Write $n - 169$ as the sum of four squares, some of which may be 0.)
24. Show that if $8 \mid a^2 + b^2 + c^2 + d^2$, then a, b, c, and d are all even.
25. Show by induction that $4^n 2$ is not the sum of four positive squares for all positive integers n.

C

The following three problems indicate a proof of Lemma 7.8 that does not use Theorem 7.6 or quadratic residue theory, and that applies to all odd primes. Assume that p *is an odd prime.*

26. Let $S = \{0^2, 1^2, 2^2, \ldots, ((p-1)/2)^2\}$. Show that no two elements of S are congruent modulo p.
27. Let $T = \{-s - 1 \colon s \text{ in } S\}$, where S is defined in the previous problem. Show that no two elements of T are congruent modulo p, but that some element x^2 of S is congruent to some element $-y^2 - 1$ of T modulo p, where x and y are nonnegative and less than $p/2$. (**Hint:** The pigeon-hole principle.)
28. Show that $x^2 + y^2 + 1 = mp$, with $0 < m < p$, where x and y are as in the previous problem.

7.4 Sums of Fourth Powers

Waring's Problem for Fourth Powers

In Section 4.3, we mentioned Waring's problem, which is to show that for each $k > 1$, there is a minimal integer $g(k)$ such that each positive integer can be written as a sum of $g(k)$ nonnegative kth powers. The 1772 theorem of Lagrange proved in the last section (along with the fact that 7 cannot be written as the sum of three squares) shows that $g(2) = 4$. Further progress on Waring's problem was not made until 1859, when Liouville showed that $g(4) \leq 53$, which is a compact way of stating the next theorem.

Theorem 7.10. *Every positive integer can be written as the sum of 53 fourth powers.*

Proof. We need the identity

$$\begin{aligned} 6(a^2+b^2+c^2+d^2)^2 &= (a+b)^4+(a-b)^4+(c+d)^4+(c-d)^4 \\ &\quad +(a+c)^4+(a-c)^4+(b+d)^4+(b-d)^4 \\ &\quad +(a+d)^4+(a-d)^4+(b+c)^4+(b-c)^4. \end{aligned} \tag{7.11}$$

Although this can be verified by multiplying out both sides, we offer a little help in cutting down the work. The left side is

$$6(a^4+b^4+c^4+d^4)+12(a^2b^2+a^2c^2+\cdots+c^2d^2).$$

Since each variable appears in six terms on the right side of (7.11), and since

$$(x \pm y)^4 = x^4 \pm 4x^3y + 6x^2y^2 \pm 4xy^3 + y^4,$$

the fourth powers match up properly. Likewise, each pair of variables (such as a and b) appears in two terms on the right side of (7.11), accounting for 12 times the product of their squares. It can be seen that the terms involving odd exponents cancel out, since each pair of variables appears once with a plus sign and once with a minus sign.

Now by Lagrange's theorem each positive integer is the sum of four squares, say $a^2+b^2+c^2+d^2$. Thus (7.11) implies that each number that is six times a square can be written as the sum of 12 fourth powers.

Let n be an arbitrary positive integer. Write

$$n = 6q + r, \quad 0 \leq r < 6.$$

Then if we write $q = w^2 + x^2 + y^2 + z^2$ (using Lagrange's theorem again), we have

$$n = 6w^2 + 6x^2 + 6y^2 + 6z^2 + r.$$

Each of the first four terms in this expression can be written as the sum of 12 fourth powers, and r, being at most 5, can be written as the sum of 5 fourth powers (of 0 and 1). Thus n is the sum of $12+12+12+12+5 = 53$ fourth powers. □

By being a little more careful we can improve the proof just given to show that $g(4) \le 50$; the details are given in the exercises. Actually, it is known that $19 \le g(4) \le 35$. That $g(4) \ge 19$ follows from the fact that 79 cannot be written as the sum of 18 fourth powers. For $3^4 = 81$ is already too big, so only the fourth powers 1 and 16 are available. Using four 16s still requires $79 - 4 \cdot 16 = 15$ ones, a total of 19 fourth powers.

Fermat's Last Theorem

The recently proved Fermat's last theorem (see Chapter 0 and Section 2.3) says that for $k > 2$, the equation

$$x^k + y^k = z^k \tag{7.12}$$

has no solution in positive integers x, y, and z. We will prove the case $k = 4$. In fact, a somewhat stronger result will be proved, namely, that

$$x^4 + y^4 = w^2 \tag{7.13}$$

has no solution in positive integers. This implies the $k = 4$ case of Fermat's last theorem, since if (7.12) had a solution with $k = 4$, then setting $w = z^2$ would give a solution to (7.13). The proof will use our analysis of Pythagorean triples in Section 7.1.

Theorem 7.11. *The equation $x^4 + y^4 = w^2$ has no solution in positive integers.*

Proof. We will start by showing that if equation (7.13) has a solution in positive integers x, y, and w, then it has another solution X, Y, W in positive integers with $W < w$.

Case 1: Some prime p divides all of x, y, and w.

Then p^4 divides $x^4 + y^4 = w^2$, from which we see that p^2 divides w. Thus

$$\left(\frac{x}{p}\right)^4 + \left(\frac{y}{p}\right)^4 = \left(\frac{w}{p^2}\right)^2,$$

and so $X = x/p$, $Y = y/p$, $W = w/p^2$ is another solution with $W < w$.

Case 2: No prime divides all of x, y, and w.

Then, since

$$(x^2)^2 + (y^2)^2 = w^2, \tag{7.14}$$

x^2, y^2, and w form a primitive Pythagorean triple. By Lemma 7.1 we conclude that x^2, y^2, and w (and so x, y, and w) are relatively prime in pairs, that exactly one of x^2 and y^2 is even, and that w is odd. Let us assume that x^2 is even, so that x is even and y is odd.

Then by the Pythagorean triple theorem applied to (7.14) there exist relatively prime positive integers u and v such that

$$x^2 = 2uv, \quad y^2 = u^2 - v^2, \quad \text{and} \quad w = u^2 + v^2.$$

We now apply the same theorem to the equation

$$y^2 + v^2 = u^2.$$

We see that y, v, and u form another primitive Pythagorean triple, and since we know that y is odd, we must have v even and u odd. Thus there exist relatively prime positive integers r and s such that

$$y = r^2 - s^2, \quad v = 2rs, \quad \text{and} \quad u = r^2 + s^2.$$

Now since $(u, v/2) = 1$, by considering the prime factorization of both sides of the equation $(x/2)^2 = u(v/2)$ we see that there must exist positive integers W and Z such that $u = W^2$ and $v/2 = Z^2$. In the same way, since $(r, s) = 1$ from the equation $rs = v/2 = Z^2$ we see that there exist positive integers X and Y such that $r = X^2$ and $s = Y^2$.

Now we have

$$X^4 + Y^4 = r^2 + s^2 = u = W^2, \tag{7.15}$$

where X, Y, and W are positive integers. Also

$$w = u^2 + v^2 = W^4 + v^2 > W^4 \geq W.$$

From equation (7.15) we see we have another solution to (7.13) with $W < w$, as claimed at the beginning of this proof. This concludes Case 2.

So far our argument has been entirely positive. We showed that given a solution x, y, w to (7.13) in positive integers, then we can find another solution X, Y, W, also in positive integers, with $W < w$. But we need not stop there. From X, Y, and W, we could create a third solution, say X', Y', W', with $W' < W$. In fact, given just one solution, we could find an infinite sequence of solutions, with

$$w > W > W' > W'' > \cdots > 0.$$

This is clearly impossible, since one cannot have an infinite decreasing sequence of positive integers. Thus no solution can exist. □

The method of proof just employed is due to Fermat, and is known as the "method of infinite descent." The argument could also be arranged as an indirect proof depending on the principle that any nonempty set of positive integers has a smallest element. Assuming a solution in positive integers to (7.13) exists, we let x, y, w be that positive solution with w minimal. Then showing the existence of the solution X, Y, Z with $W < w$ provides a contradiction. Since the principle that any nonempty set of positive integers has a least element (called the "well-ordering principle") can be shown to be equivalent to mathematical induction, infinite descent can be considered to be a variation on induction. See the problems at the end of Sections 1.6 and 2.1 for the equivalence of the principles of mathematical induction and well-ordering.

Problems for Section 7.4.

A

In the first six problems, write n as the sum of as few positive cubes as possible.

1. $n = 20$
2. $n = 30$
3. $n = 40$
4. $n = 52$
5. $n = 100$
6. $n = 155$

In the next four problems, write n as the sum of as few positive fourth powers as possible.

7. $n = 100$
8. $n = 177$
9. $n = 200$
10. $n = 300$

11. Find the smallest positive integer that cannot be written as the sum of eight nonnegative cubes.
12. Find the second smallest positive integer that cannot be written as the sum of 18 fourth powers.

B

13. Show that if $0 < n < 79$, then n can be written as the sum of 18 fourth powers.
14. Show that if $n > 80$, then $n = 6q + r$, where $r = 0, 1, 2, 16, 17$, or 81.
15. Show that $g(4) \leq 50$.
16. Let k be a positive integer, and define $t = \lfloor (3/2)^k \rfloor$. Show that $2^k t - 1$ cannot be expressed as the sum of fewer than $2^k + t - 2$ nonnegative kth powers. (**Hint:** Show $2^k t - 1 < 3^k$.)
17. Show that $x^4 \equiv 0$ or $1 \pmod{16}$ for all integers x. Conclude that no number congruent to 15 (mod 16) can be written as the sum of fewer than 15 fourth powers.
18. Show that if $n \equiv 4$ or $5 \pmod 9$, then n cannot be written as the sum of three nonnegative cubes.
19. Show that if n is any positive integer then $x^{4n} + y^{4n} = z^{4n}$ is impossible in positive integers.
20. Show that to prove Fermat's last theorem it suffices to show that if p is any odd prime, then $x^p + y^p = z^p$ is impossible in positive integers.
21. Show that $x^4 - 4y^4 = w^2$ has no solution in positive integers. (**Hint:** Compute $w^4 + (2xy)^4$.)
22. Show that given any positive integer n there exists a positive integer a such that $n + a^2 < (a+1)^2$.

C

23. One way to represent a positive integer n as a sum of squares is to let $x_1 = \lfloor\sqrt{n}\rfloor$, so $x_1^2 \le n$. Then let $n_1 = n - x_1^2$. If $n_1 > 0$, let $x_2 = \lfloor\sqrt{n_1}\rfloor$, and $n_2 = n_1 - x_2^2$. Eventually we get $n_{k+1} = 0$, and $n = x_1^2 + x_2^2 + \cdots + x_k^2$. (This is called the "greedy algorithm," since at each step we take as big a bite out of n as possible.) Show that given k there exists n such that this algorithm represents n as the sum of k nonzero squares. (**Hint:** See the previous exercise.)

In the following four problems, assume $x^4 - y^4 = w^2$ has a solution in positive integers, and the x, y, w is such a solution with x minimal.

24. Show that x, y, and w are relatively prime in pairs.
25. Show that x is odd.
26. Show that if y is odd there exist relatively prime positive integers u and v with $(xy)^2 = u^4 - v^4$ and $u < x$.
27. Show that if y is even, there exist relatively prime positive integers u and v with v even such that $y^2 = 2uv$ and $x^2 = u^2 + v^2$. Then show there exist relatively prime positive integers a and b such that $u = a^2$, $v/2 = b^2$, and $x^2 = a^4 + 4b^4$. Show that there exist relatively prime positive integers U and V such that $a^2 = U^2 - V^2$ and $b^2 = UV$. Then show there exist positive integers c and d such that $U = c^2$, $V = d^2$, $a^2 = c^4 - d^4$, and $c < x$.
28. Show that the equation $x^4 - y^4 = w^2$ has no solutions in positive integers.

7.5 Pell's Equation

A Misnamed Equation

The equation $x^2 - dy^2 = c$, where d and c are given constants and x and y are the unknowns, is known as **Pell's equation**, due to a mistake by Euler. The title is unfortunate, since the English mathematician John Pell (1610–1685) made no contribution to the theory of the equation. If $d < 0$, clearly x and y cannot be too large; and if solutions exist, there are only a finite number, which can be found by trial and error.

Likewise if d is a perfect square, say $d = a^2$, then the equation can be written as

$$(x + ay)(x - ay) = c$$

and all solutions may be found by considering the possible factorizations of c.

Example. Find all nonnegative solutions to $x^2 - 9y^2 = -161$.

The equation can be written as $(x + 3y)(x - 3y) = -161$. Since $161 = 7 \cdot 23$, and since $x + 3y$ cannot be negative, the possibilities for $x + 3y$ and $x - 3y$ are 161 and -1, 1 and -161, 23 and -7, and 7 and -23. Only the first and third cases lead to nonnegative integral values for x and y, namely, $x = 80$ and $y = 27$, and $x = 8$ and $y = 5$. ◊

From now on we will concentrate on the case of Pell's equation when infinitely many solutions are at least a possibility, namely, when $d > 0$ and d is not a perfect square. Then $\sqrt{d}$ is irrational by Theorem 6.1. Clearly it suffices to look for nonnegative solutions; the other solutions can easily be generated from these. We will usually consider cases when c is small compared to d, and our strongest result will be for $c = 1$. Then x^2 will have to be close to dy^2 for x, y to be a solution, and so, dividing by y^2, $(x/y)^2$ will be close to d. This means that the rational x/y will be close to the irrational number $\sqrt{d}$.

When investigating infinite continued fractions we mentioned that the convergents to an irrational number are good rational approximations to that number. The next theorem makes that statement somewhat more definite.

Theorem 7.12. *Suppose the positive integer d is not a perfect square, and let $[q_1, q_2, \dots]$ be the continued fraction expansion for $\sqrt{d}$. Let $h_i/k_i = [q_1, \dots, q_i]$ be the ith convergent to $\sqrt{d}$, where the integers h_i and k_i are defined as in Section 6.3. Then*

$$|h_i^2 - dk_i^2| < 2\sqrt{d} + 1$$

for all $i \geq 0$.

Proof. The inequality is clear for $i = 0$ since $h_0 = 1$ and $k_0 = 0$. By Lemma 6.12 and Theorems 6.14 and 6.15 we have

$$\frac{h_1}{k_1} < \frac{h_3}{k_3} < \cdots < \sqrt{d} < \cdots < \frac{h_4}{k_4} < \frac{h_2}{k_2}.$$

Thus for any $i > 1$, we have

$$\left|\frac{h_{i-1}}{k_{i-1}} - \sqrt{d}\right| < \left|\frac{h_{i-1}}{k_{i-1}} - \frac{h_i}{k_i}\right|.$$

By (6.5) (in Section 6.3) the latter quantity equals $1/k_i k_{i-1}$. Then $|h_{i-1} - k_{i-1}\sqrt{d}| < 1/k_i$, and so $h_{i-1} = k_{i-1}\sqrt{d} + E$, where $|E| < 1/k_i$. Thus

$$h_{i-1}^2 = k_{i-1}^2 d + 2k_{i-1}\sqrt{d}E + E^2,$$

and

$$|h_{i-1}^2 - k_{i-1}^2 d| = |2k_{i-1}\sqrt{d}E + E^2| < \frac{2k_{i-1}\sqrt{d}}{k_i} + \frac{1}{k_i^2} \leq 2\sqrt{d} + 1,$$

since the numbers k_i form a nondecreasing sequence of positive integers for $i > 0$. □

Example. If $d = 3$, we compute $\sqrt{3} = [1, 1, 2, 1, 2, \ldots] = [1, \overline{1, 2}]$ by the methods of Section 6.3. Notice that all the values of $h_i^2 - 3k_i^2$ below are less in absolute value than $2\sqrt{3} + 1$, which is about 4.46.

i	h_i	k_i	q_i	$h_i^2 - 3k_i^2$
−1	0	1		
0	1	0		
1	1	1	1	$1 - 3 = -2$
2	2	1	1	$4 - 3 = 1$
3	5	3	2	$25 - 3 \cdot 9 = -2$
4	7	4	1	$49 - 3 \cdot 16 = 1$
5	19	11	2	$361 - 3 \cdot 121 = -2$

Now we consider the equation $x^2 - dy^2 = 1$, where $d > 0$ is not a perfect square. Clearly, x cannot be 0, and if $y = 0$ then x must be 1 or −1. Thus we will concentrate on solutions where both x and y are positive.

Theorem 7.13. *If the positive integer d is not a perfect square, then $x^2 - dy^2 = 1$ has a solution in positive integers x and y.*

Proof. By the previous theorem there are infinitely many pairs of relatively prime positive integers x and y such that $|x^2 - dy^2| < 2\sqrt{d} + 1$, namely, the pairs h_i, k_i corresponding to the continued fraction expansion of $\sqrt{d}$. Thus there must exist an integer $t \neq 0$ such that $x^2 - dy^2 = t$ for infinitely many such x and y. (We cannot have $x^2 - dy^2 = 0$ for positive x and y since then we would have $\sqrt{d} = x/y$, a rational number.)

If r is the least residue of $x \pmod{|t|}$ and s is the least residue of $y \pmod{|t|}$, then the pair (r, s) can assume only t^2 values. Thus there exist two pairs x, y and X, Y such that

$$x^2 - dy^2 = t \quad \text{and} \quad X^2 - dY^2 = t,$$

with

$$x \equiv X \pmod{|t|}, \quad \text{and} \quad y \equiv Y \pmod{|t|}.$$

Note that

$$xX - yYd \equiv x^2 - y^2 d = t \equiv 0 \pmod{|t|},$$

and

$$xY - Xy \equiv xy - xy = 0 \pmod{|t|},$$

so there exist integers u and v such that

$$xX - yYd = tu \quad \text{and} \quad xY - Xy = tv.$$

Then

$$\begin{aligned}
t^2 &= (x^2 - dy^2)(X^2 - dY^2) \\
&= (x - y\sqrt{d})(x + y\sqrt{d})(X - Y\sqrt{d})(X + Y\sqrt{d}) \\
&= (x - y\sqrt{d})(X + Y\sqrt{d})(x + y\sqrt{d})(X - Y\sqrt{d}) \\
&= (xX - yYd + (xY - Xy)\sqrt{d})(xX - yYd - (xY - Xy)\sqrt{d}) \\
&= (tu + tv\sqrt{d})(tu - tv\sqrt{d}) = t^2(u^2 - dv^2),
\end{aligned}$$

and so $u^2 - dv^2 = 1$. The desired solution is the pair $|u|, |v|$. It remains to show that $v \neq 0$. But if $v = 0$, then $xY = Xy$, and $x/y = X/Y$. But we know the convergents to an irrational number are distinct. □

Infinitely Many Solutions

Although we will not prove it here, a solution to $x^2 - dy^2 = 1$ will always be found among the pairs h_i, k_i corresponding to the continued fraction expansion of $\sqrt{d}$. This is important because the next theorem shows how to generate infinitely many solutions to this equation from a single one, and then how to generate infinitely many solutions to the more general equation

$$x^2 - dy^2 = c,$$

given one solution.

Theorem 7.14. *Suppose the positive integer d is not a perfect square.*

1. *If x, y is any positive solution to*

$$x^2 - dy^2 = 1, \tag{7.16}$$

and if $n > 0$, then the pairs of integers x_n, y_n uniquely defined by $x_n + y_n\sqrt{d} = (x + y\sqrt{d})^n$ are also a positive solution to (7.16), and these are distinct for distinct values of n.

2. *If u, v is any positive solution to*

$$u^2 - dv^2 = c, \tag{7.17}$$

and if x, y is any of the infinitely many positive solutions to (7.16), then the pair U, V, defined by

$$U = ux + vyd, \quad V = uy + vx,$$

is also a positive solution to (7.17), and these are distinct for distinct positive solutions x, y.

Proof. (1) The first question is whether x_n and y_n are uniquely defined. Expanding $(x + y\sqrt{d})^n$ gives an expression of the form $X + Y\sqrt{d}$, where X and Y are positive integers. Suppose $X + Y\sqrt{d} = U + V\sqrt{d}$ for any integers X, Y, U, and V. Then $(Y - V)\sqrt{d} = U - X$. If $Y - V$ is not 0, then this equation implies that $\sqrt{d}$ is rational. Thus $V = Y$ and so $U = X$.

Notice that if instead we expand $(x - y\sqrt{d})^n$, all the odd powers of $\sqrt{d}$ enter with a minus sign and we get $x_n - y_n\sqrt{d}$. Thus

$$\begin{aligned} x_n^2 - dy_n^2 &= (x_n - y_n\sqrt{d})(x_n + y_n\sqrt{d}) \\ &= (x - y\sqrt{d})^n(x + y\sqrt{d})^n = (x^2 - dy^2)^n = 1. \end{aligned}$$

We see that x_n, y_n is a solution for each n.

Since $x + y\sqrt{d} > 1$ the numbers $x_n + y_n\sqrt{d} = (x + y\sqrt{d})^n$ form a strictly increasing sequence. Thus the pairs x_n, y_n must be distinct. In fact, since $x_n^2 - dy_n^2 = 1$, both sequences x_n and y_n increase.

(2) From equations (7.16) and (7.17) we have

$$\begin{aligned} U^2 - dV^2 &= u^2x^2 + 2uvxyd + v^2y^2d^2 - d(u^2y^2 + 2uvxy + v^2x^2) \\ &= u^2(x^2 - dy^2) - dv^2(x^2 - dy^2) = u^2 - dv^2 = c. \end{aligned}$$

If x, y and x', y' are two solutions to (7.16) with $x > x'$, then $y > y'$ also. Then $U' > U$ and $V' > V$, where U' and V' are defined in the obvious way. □

Example. Find two more positive solutions to $x^2 - 7y^2 = 1$ besides $x = 8$, $y = 3$.

We have $(8 + 3\sqrt{7})^2 = 127 + 48\sqrt{7}$, so $x = 127, y = 48$ is another solution. Likewise $(8+3\sqrt{7})^3 = (8+3\sqrt{7})(127+48\sqrt{7}) = 2024+765\sqrt{7}$, so $x = 2024, y = 765$ is a third solution. ♢

Example. Find another positive solution to $x^2 - 7y^2 = -3$ besides $x = 5, y = 2$.

We take $u = 5, v = 2$, and, from the previous example, $x = 8, y = 3$. Then $U = 5 \cdot 8 + 2 \cdot 3 \cdot 7 = 82$, $V = 5 \cdot 3 + 2 \cdot 8 = 31$ satisfies $U^2 - 7V^2 = -3$. ♢

The value $c = 1$ is the only one for which Pell's equation is always assured of having a solution. Although we used continued fractions to show the existence of solutions in this case, we have not indicated the full connection between solutions of $x^2 - dy^2 = c$ and the continued fraction expansion of $\sqrt{d}$. For example, it can be shown that if $|c| < d$, then any positive solution to this equation corresponds to the numerator and denominator of a convergent h_i/k_i. Since the quadratic irrational $\sqrt{d}$ has a periodic continued fraction expansion, the question of for which values of c with $|c| < d$ the equation has solutions, and the form of these solutions, is reduced to a finite determination. We have not gone into the theory of the continued fraction expansions of quadratic irrationalities sufficiently to give details. The reader is referred to the book by Niven, Zuckerman, and Montgomery listed in the references for a fuller treatment.

DEFINITION. minimal solution

Suppose $d > 0$ is not a perfect square. We say that x and y satisfying $X^2 - dY^2 = c$ is a **minimal solution** if x and y are positive, and if whenever x', y' is any other positive solution, then $x' > x$. There can be only one minimal solution.

It turns out that if x, y is the minimal solution to $X^2 - dY^2 = 1$, then all positive solutions are generated as in part (1) of the last theorem. Also if x, y is the minimal solution to $X^2 - dY^2 = -1$, then u, v is the minimal solution to $X^2 - dY^2 = 1$, where $u + v\sqrt{d} = (x + y\sqrt{d})^2$. Proofs of these results will be found in the exercises.

Problems for Section 7.5.

A

In the first six problems, use part (1) of Theorem 7.14 to find two more positive solutions to the the equation $X^2 - dY^2 = 1$, given that x, y is a solution.

1. $d = 2, x = 3, y = 2$
2. $d = 15, x = 4, y = 1$
3. $d = 6, x = 5, y = 2$
4. $d = 3, x = 7, y = 4$
5. $d = 11, x = 10, y = 3$
6. $d = 63, x = 8, y = 1$

In the next four problems, compute $h_i^2 - dk_i^2$ for $i = 1, 2, 3,$ and 4, where h_i/k_i is the ith convergent to $\sqrt{d}$.

7. $d = 7$
8. $d = 13$
9. $d = 12$
10. $d = 10$

In the next four problems, find the smallest two solutions x, y to the given equation where x/y is a convergent to $\sqrt{d}$.

11. $x^2 - 8y^2 = -4 \quad (\sqrt{8} = [2, \overline{1, 4}])$
12. $x^2 - 6y^2 = 1 \quad (\sqrt{6} = [2, \overline{2, 4}])$
13. $x^2 - 5y^2 = 1 \quad (\sqrt{5} = [2, \overline{4}])$
14. $x^2 - 11y^2 = -2 \quad (\sqrt{11} = [3, \overline{3, 6}])$

In the next four problems, find by inspection the minimal solutions to $x^2 - dy^2 = 1$ and $x^2 - dy^2 = c$, then use the method of part (2) of Theorem 7.14 to find another solution to the second equation.

15. $d = 24, c = -20$
16. $d = 6, c = 3$
17. $d = 5, c = -19$
18. $d = 30, c = 19$

In the next six problems, find all nonnegative solutions to the given equation.

19. $x^2 + 3y^2 = 28$
20. $x^2 + 5y^2 = 69$
21. $x^2 - 4y^2 = 45$
22. $x^2 - 9y^2 = 64$
23. $x^2 - 16y^2 = -23$
24. $x^2 - 25y^2 = -56$

B

In the remaining problems, assume that $d > 0$ is not a perfect square.

25. Let X, Y be a positive solution to $x^2 - dy^2 = -1$. Show that if $n > 0$, there exist unique integers x_n and y_n such that $x_n + y_n\sqrt{d} = (X + Y\sqrt{d})^n$, and that $x_n^2 - dy_n^2 = -1$ if and only if n is odd.
26. Show that if X, Y is a positive solution to $x^2 - dy^2 = 1$, then for any integer n (positive or not) there exist unique integers x_n and y_n such that $x_n + y_n\sqrt{d} = (X + Y\sqrt{d})^n$, and that $x_n^2 - dy_n^2 = 1$ also.
27. Show that if $x^2 - dy^2 = -1$ is solvable, then d is congruent to 1 or 2 (mod 4).
28. Show that if $x^2 - dy^2 = -1$ is solvable, then no prime congruent to 3 (mod 4) divides d.

C

In the next four problems, assume that x, y is a minimal solution to $r^2 - ds^2 = 1$, and that x_n and y_n are defined as in part (1) of Theorem 7.14.

29. Show that if $X^2 - dY^2 = 1$ with X and Y positive and different from the pair x, y, then $x + y\sqrt{d} < X + Y\sqrt{d}$.
30. Suppose that X, Y is a positive solution to $r^2 - ds^2 = 1$, but that X, Y is not equal to any pair x_n, y_n, $n = 0, 1, \ldots$. Show that there exists $m > 0$ such that $(x + y\sqrt{d})^m < X + Y\sqrt{d} < (x + y\sqrt{d})^{m+1}$.
31. Assume the hypothesis of the last problem. Let $u = Xx_m - Yy_m d$ and $v = Yx_m - Xy_m$. Show that $u^2 - dv^2 = 1$, $1 < u + v\sqrt{d} < x + y\sqrt{d}$, $0 < u - v\sqrt{d} < 1$, and that u and v are positive integers.
32. Show that every solution in positive integers to $r^2 - ds^2 = 1$ is one of the pairs x_n, y_n for some $n > 0$.

In the next five problems, suppose x, y is the minimal solution to $X^2 - dY^2 = 1$, and u, v is the minimal solution to $X^2 - dY^2 = -1$. Let $r = u^2 + dv^2$ and $s = 2uv$.

33. Show that $r^2 - ds^2 = 1$, $0 < -u + v\sqrt{d} < 1$, and $r + s\sqrt{d} \geq x + y\sqrt{d}$.
34. Suppose $r \neq x$. Let $h = -xu + yvd$, $k = xv - yu$. Show that $h^2 - dk^2 = -1$ and $-u + v\sqrt{d} < h + k\sqrt{d} < u + v\sqrt{d}$.
35. Given the hypotheses of the previous problem, show that $hk < 0$.
36. Given the hypotheses of problem 34, show that $hk < 0$ is impossible. (**Hint:** If $h < 0$ and $k > 0$ show that $u + v\sqrt{d} > -h + k\sqrt{d}$. If $h > 0$ and $k < 0$, derive a contradiction from the definitions of h and k.)
37. Show that $x = r$ and $y = s$.

Bibliography

ANDREWS, GEORGE E. *Number Theory.* New York: Dover, 1998.

BELL, ERIC T. *Men of Mathematics.* New York: Simon and Schuster, 1986.

BLUM, MANUEL. "Coin Flipping by Telephone—a Protocol for Solving Impossible Problems." *Proc. IEEE Spring Comp. Conf.*, 1982, 133–137.

BOYER, CARL B. *A History of Mathematics*, 2nd ed. New York: Wiley, 1991.

BRILLHART, JOHN, D. H. LEHMER, J. L. SELFRIDGE, BRYANT TUCKERMAN, and S. S. WAGSTAFF, JR. *Factorizations of* $b^n \pm 1$, $= 2, 3, 5, 6, 7, 10, 11, 12$ *up to Higher Powers*, 2nd ed. Providence, RI: Am. Math. Soc., 1988.

BURTON, DAVID M. *Elementary Number Theory*, 4th ed. New York: McGraw-Hill, 1997.

BURTON, DAVID M. *The History of Mathematics*, 4th ed. New York: McGraw-Hill, 1998.

CAJORI, FLORIAN. *A History of Mathematics*, 5th ed. Providence, RI: Am. Math. Soc., 1999.

DAVENPORT, HAROLD. *The Higher Arithmetic*, 7th ed. New York: Cambridge, 1998.

DEDRON, P., and J. ITARD. *Mathematics and Mathematicians*, 2 vols. London: Transworld, 1974.

DICKSON, L. E. *History of the Theory of Numbers*, 3 vols. New York: Chelsea, 1952.

DUDLEY, UNDERWOOD. *Elementary Number Theory*, 2nd ed. New York: W. H. Freeman, 1978.

EDGAR, HUGH M. *A First Course in Number Theory.* Boston: PWS Publishers, 1987.

EDWARDS, HAROLD M. *Fermat's Last Theorem: A Genetic Approach to Algebraic Number Theory.* New York: Springer-Verlag, 1996.

EUCLID. *The Elements*, 3 vols. New York: Dover, 1956.

EVES, HOWARD W. *An Introduction to the History of Mathematics*, 6th ed. Philadelphia: Saunders, 1990.

FIELD, LEEDS K. *Mathematics, Minus and Plus.* Memphis: Pageant Press, 1953.

GROSSWALD, EMIL. *Topics from the Theory of Numbers*, Cambridge, MA: Birkhäuser, 1984.

HALBERSTAM, H., and H. E. RICHERT. *Sieve Methods.* San Diego: Academic Press, 1975.

HARDY, G. H. *A Mathematician's Apology.* New York: Cambridge, 1992.

HARDY, G. H., and E. M. WRIGHT. *An Introduction to the Theory of Numbers*, 5th ed. New York: Oxford, 1980.

HELLMAN, MARTIN E. "The Mathematics of Public-Key Cryptography," *Scientific American*, August 1979, 146–157.

HOFFMAN, PAUL. *The Man Who Loved Only Numbers: The Story of Paul Erdös and the Search for Mathematical Truth.* New York: Hyperion, 1998.

HUA, L. K. *Introduction to Number Theory.* New York: Springer-Verlag, 1987.

LANDAU, EDMUND. *Elementary Number Theory.* New York: Chelsea, 1958.

LEVEQUE, WILLIAM J. *Fundamentals of Number Theory.* New York: Dover, 1996.

LONG, CALVIN T. *Elementary Introduction to Number Theory*, 3rd ed. Prospect Heights, IL: Waveland Press, 1995.

NAGELL, TRYGVE. *Introduction to Number Theory.* Jamaica, NY: Chelsea, 1981.

NEUGEBAUER, OTTO. *The Exact Sciences in Antiquity*, 2nd ed. New York: Dover, 1969.

NIVEN, IVAN, HERBERT S. ZUCKERMAN, and HUGH L. MONTGOMERY. *An Introduction to the Theory of Numbers*, 5th ed. New York: Wiley, 1991.

ORE, OYSTEIN. *Number Theory and Its History.* New York: Dover, 1988.

POMERANCE, CARL. "The Search for Prime Numbers," *Scientific American*, December 1982, 136–147.

RIESEL, HANS. *Prime Numbers and Computer Methods for Factorization.* Boston: Birkhäuser, 1985.

ROBINSON, RAPHAEL M. "Mersenne and Fermat Numbers," *Proc. Amer. Math. Soc.*, 1954, 842–846.

ROSEN, KENNETH H. *Elementary Number Theory and Its Applications*, 4th ed. Reading, MA: Addison-Wesley, 1999.

SCHECHTER, BRUCE. *My Brain Is Open: The Mathematical Journeys of Paul Erdös.* New York: Simon & Schuster, 1998.

SCHROEDER, M. R. *Number Theory in Science and Communication*, 3rd ed. New York: Springer-Verlag, 1997.

SHANKS, DANIEL. *Solved and Unsolved Problems in Number Theory*, 4th ed. New York: Chelsea, 1985.

SMITH, DAVID EUGENE. *Portraits of Eminent Mathematicians*, 2 vols. New York: Scripta Mathematica, 1936, 1938.

STARK, HAROLD M. *An Introduction to Number Theory.* Cambridge, MA: MIT Press, 1978.

VANDEN EYNDEN, CHARLES. *Number Theory: An Introduction to Proof.* Scranton, PA: Intext, 1970.

Answers to Odd-Numbered Problems

Chapter 0

1. 3, 5, 11, 17, 29, 41   3. 10   7. 127   9. 28
13. 7, 13, 19, 31, 37, 43, 61, 67, 73, 79, 97   15. none   17. 4

Chapter 1

Section 1.1

1. T, F, F   3. 9   5. 17, 102   7. 18, 36, 54, 72, 108, 216
9. $\pm1, \pm2, \pm3, \pm4, \pm6, \pm8, \pm12, \pm24$ 11. $0, \pm4, \pm8, \pm12, \pm16, \pm20, \pm24$
13.

b	1	2	3	4	5	6	7	8	9
(a,b)	1	2	1	4	1	2	1	8	1
$[a,b]$	8	8	24	8	40	24	56	8	72
$(a,b)[a,b]$	8	16	24	32	40	48	56	64	72

15. all   17. ±1 19. T   21. $a = 1$, $b = c = 2$ 23. T   25. T
27. T

Section 1.2

1. 29, 3   3. 0, 155   5. 3, 0   7, 0, 0   9. -5, 5   11. -12, 16
13. $\pm1, \pm2, \pm3, \pm6$ 15. all multiples of 12
17. $b = 1$, $c = -1$ 19. T   21. T   23. T   25. $a = 4$, $b = 8$
27. $a = 9$, $b = 672$

Section 1.3

1. 31; $x = -3$, $y = 2$ 3. 11; $x = 19$, $y = -6$ 5. 1; $x = -21$, $y = 34$
7. 29; $x = 9$, $y = 4$ 9. $x = 6$, $y = -11$ 11. $x = -6$, $y = 4$
13.

x	1	4	7	10	13	16	19	22
y	16	14	12	10	8	6	4	2

15.

x	3	6	9	12
y	8	6	4	2

17. Because $2 \mid 4x + 6y$ for any integers x and y. 19. $x = 2 + 3k$
27. By our definition 0 has no multiples.

Section 1.4

1. yes 3. no 5. yes 7. $x = 9$, $y = -3$ 9. $x = 700$, $y = -900$
11. $x = -4$, $y = -4$

13.

x	2	8	14
y	15	10	5

15.

x	4	8	12	16
y	12	9	6	3

17.

x	21	72
y	19	8

19.

x	28	66
y	31	8

21. Those such that $b \mid a$ and $b \mid c$ or $a = b = c = 0$. 23. 11 cents
25. 6 cows, 10 pigs 27. impossible 29. 52 cents
31. 17 pens and 45 notebooks 33. 19 cents 35. $x = -109$, $y = -111$

Section 1.5

1. T 3. F 5. F 7. F 9. 2 11. 17 13. 38 15. 5
17. 209 19. 19880 21. 25407 23. 47 25. 6 27. 15 29. 5
31. 1 33. $3x \equiv 7y \pmod 5$ 35. $2x \equiv 5y \pmod 5$ 37. $a = 2$
39. $a = c = 2$, $b = 1$ 41. $a = d = 2$, $b = c = 1$

Section 1.6

1. $A = 2$, $R = 2$, $n = 5$; 62 3. no 5. no
7. $A = -1/2$, $R = -2$, $n = 3$; $-3/2$ 9. no
11. $A = 1$, $R = 1/3$, $n = 6$; 364/243
13. $A = 3/2$, $R = 4/9$, $n = 4$; 1261/486 15. 1597, 2584, 4181, 6765
59. Induction condition 2 doesn't hold when $k = 1$.

Chapter 2

Section 2.1

1. 23, 29, 31, 37 3. 101, 103, 107, 109, 113

5.

n	11	12	13	14	15	16	17	18	19	20
$\tau(n)$	2	6	2	4	4	5	2	6	2	6

7. $2 \cdot 2 \cdot 3 \cdot 3 \cdot 5$
9. $7 \cdot 11 \cdot 13$ 11. 6 13. 6 15. 2 17. 6 19. $n + 1$ 21. 16
23. 45360

Section 2.2

1. 2, 6, 10, 14, 18 3. 4, 7, 10, 13, 19, 22, 25 5. 4 7. 0 9. 4
11. 4 13. 0 15. $100 = 10 \cdot 10 = 4 \cdot 25$ 17. $6^2 = 2 \cdot 18$

Section 2.3

1. 103 3. $7 + 3\sqrt{-6}$ 5. $-7 - 2\sqrt{-6}$ 7. $11 + \sqrt{-6}$ 9. $5 + 3\sqrt{-6}$
13. no; ± 1, $1 \pm \sqrt{-6}$, $-1 \pm \sqrt{-6}$, ± 7 21. T 23. T 25. T 27. T

29. $p = a = b = 2$ 31. $p = a = 2$, $b = -2$
33. The calculator must be wrong since $5 \nmid 2^{33}$ and $2 \nmid 8589934605$. 41. 84

Section 2.4

1. 293^1 3. $41^1 43^1$ 5. 3^8 7. $11^1 101^1$ 9. 11^4 11. 3^1 13. 3^0
15. no, 99, 19305 17. no, 2, 99484 19. no, 1, 3993990
21. when k is even or $d = 0$ 23. when $9 \mid kd$ 27. 2, 1 29. 1, 4
31. 1, 10 33. 8 35. 44 39. 12
41. $3 \mid n$ if and only if 3 divides the last digit of n

Section 2.5

1. 53, 59, 61, 67, 71, 73, 79, 83, 89, 97 3. 1009, 1013, 1019, 1021
5.

n	11	12	13	14	15	16	17	18	19	20
(P, n)	11	6	13	14	15	2	17	6	19	10

7. Let $p_1, p_2, \ldots, p_k$ be the distinct primes dividing n. Then $P(n) = p_1 p_2 \cdots p_k$.
9. 0, 1 15. $2 \cdot 3 \cdot 5 \cdot 7 \cdot 11 \cdot 13 + 1 = 59 \cdot 509$

Section 2.6

1. 6 3. 8 5. 49 7. 27 9. 6
11.

	1	2	4	8
1	1	2	4	8
3	3	6	12	24
5	5	10	20	40
15	15	30	60	120

yes, yes, yes
13.

	1	2	4	7	14	28
1	1	2	4	7	14	28
3	3	6	12	21	42	84
7	7	14	28	49	98	196
21	21	42	84	147	294	588

yes, yes, no
17. no 19. yes 21. yes 23. $f(n) = g(n) = 1$ for all n, $a = b = 1$
25. $f(n) = 1$ for all n, $C = 2$, $a = b = 1$ 27. T 29. $f(n) = 0$ for all n
31. 24 33. 129600 35. 8 51. 18

Chapter 3

Section 3.1

1. 42 3. 217 5. 102 7. 720 9. 558 11. 50 13. 4550
15. 8/7 17. 104/63 19. 55 21. 31 23. 6 25. 31 27. 1/2
29. 396

Section 3.2

1.

n	31	32	33	34	35	36	37	38	39	40
$\sigma(n)$	32	63	48	54	48	91	38	60	56	90
n	41	42	43	44	45	46	47	48	49	50
$\sigma(n)$	42	96	44	84	78	72	48	124	57	93

3. 12

5.

n	1	2	3	4	5	6	7	8	9	10
$G(n)$	1	1	2	1	2	2	2	1	3	2

7.

n	1	2	3	4	5	6	7	8	9	10
$M(n)$	1	-1	-1	0	-1	1	-1	0	0	1

11. $f(1) = 1$, $f(3) = 5/4$ 17. 1, 3, 6, 9

19. $F(n) = \prod_{i=1}^{k} \frac{(m_i + 1)(m_i + 2)}{2}$

21. $F(n) = \prod_{\substack{p \mid n \\ p \text{ prime}}} (-p)$

Section 3.3

1. 8128 3. $2^9 - 1 = 7 \cdot 73$ 5. (a) $3 \cdot 5$ (b) 3 7. (a) $3 \cdot 5 \cdot 17$ (b) 3, 15
9. (a) $3 \cdot 11 \cdot 31$ (b) 3, 31 11. (a) 3^2 (b) 0 (c) 3 13. (a) $5 \cdot 13$ (b) 1 (c) 5
15. (a) $3^3 \cdot 19$ (b) 0 (c) 3 17. To have $2^{r+1} - 1 > 1$ we need $r > 0$.

Section 3.4

1. $3 \cdot 683$ 3. $3 \cdot 2731$ 5. $3^2 \cdot 11 \cdot 331$ 7. F, $k = 4$ 9. N
11. F, $k = 0$ 13. T

Section 3.5

1. 1 ~~2~~ 3 ~~4~~ ; 2
3. 1 ~~2~~ 3 ~~4~~ 5 ~~6~~ 7 ~~8~~ 9 ~~10~~ 11 ~~12~~ 13 ~~14~~ 15 ~~16~~ ; 8
5. 1 2 ~~3~~ ; 2
7. 1 2 ~~3~~ 4 ~~5~~ ~~6~~ 7 8 ~~9~~ ~~10~~ 11 ~~12~~ 13 14 ~~15~~ ; 8
9. yes 11. yes 13. no 15. no 17. no 19. no 21. 40
23. 18 25. 648 27. 72 29. (a) 10 (b) 6 (c) 2 (d) $10 - 2 = 8$
31. $\phi(1) + \phi(2) + \phi(3) + \phi(4) + \phi(6) + \phi(12) = 1 + 1 + 2 + 2 + 2 + 4 = 12$
35. $x = 1, 2$ 37. $x = 5, 8, 10, 12$ 43. odd n 55. 100^8 works

Section 3.6

1.

n	1	2	3	4	5	6	7	8	9	10	11	12
$\mu(n)$	1	-1	-1	0	-1	1	-1	0	0	1	-1	0

3. 26, 33, 34, 35, 38, 39, 46 5. 30, 42, 66, 70, 78
7. $\mu(1) + \mu(2) + \mu(3) + \mu(4) + \mu(6) + \mu(12) = 1 - 1 - 1 + 0 + 1 + 0 = 0$
9. If $p^3 \mid n$ for any prime p, then $f(n) = 0$. Otherwise $f(n) = \prod_{\substack{p \| n \\ p \text{ prime}}} (-2)$.

11. $f(1) = 1$, $f(2) = -1$, and $f(n) = 0$ otherwise
21. $f(n) = \prod_{\substack{p^k \| n \\ p \text{ prime}}} (-1)$

Chapter 4

Section 4.1

1. no 3. yes 5. no 7. no 9. no 11. yes 13. no
15. no 17. 5 19. 13 21. 11 23. 0 25. 3, 7
27. 4, 11, 18, 25, 32 29. 1, 6, 11 31. 0, 6, 12 33. 109
35. 1, 3, 5, 7, 9 37. 102, 107 43. $b = c = 2$, $x_1 = 0$, $x_2 = 1$ 45. T
47. $b = 2$

Section 4.2

1. $z = 47$ 3. $x = 20, 119$ 5. $x = 2, 107$ 7. $x = 61$ 9. $z = 199$
11. $q = 19, 327$ 13. no 15. yes 17. no 19. no solutions
31. $b = 2$, $x_1 = 1$ 33. $b = 2$, $a = x_1 = 1$ 35. T 37. 1, 3, 7, 29
39. 2020 41. 7 A.M. Friday, August 1
43. every 35 days; August 7, 10 P.M. 45. 463

Section 4.3

1. 1, 4, 4, 2 3. 1, 10, 5, 5, 5, 10, 10, 10, 5, 2 5. 4; 8 7. 3; 6
9. 18; 18 11. 10 13. 7 15. 1 17. 0 19. 07 21. 0
23. 100
27. $16! = (2 \cdot 9)(3 \cdot 6)(4 \cdot 13)(5 \cdot 7)(8 \cdot 15)(10 \cdot 12)(11 \cdot 14)16 \equiv -1 \pmod{17}$
29. 1 33. $k = 4$, $b = 8$

Section 4.4

1. 110101 3. 1101111 5. 29 7. 830 9. 4 11. 28
13. $2^{90} \equiv 64 \pmod{91}$
15. 4, 14, 194, 788, 701, 119, 1877, 240, 282, 1736; M_{11} is not prime
17. $3^8 \equiv -1 \pmod{17}$

Section 4.5

1. LECC GF DGFBNX 3. MHEKMHEZ CN REDDE 5. ABSTAIN FROM BEANS
7. THE CLOCK WILL EXPLODE
9. NVNQY DKTNBNQ DS TCN SUJ ME EMUQ SPULQNS, THE THEOREM OF WILSON IS TRUE
11. 12, 20, 10, 311 13. 70, 520, 0, 112 15. 331, 877 17. 509, 535
19. 160, 520 21. 406, 930 23. $b = 30$, $d = 13$
25. $b = 198$, $d = 41$ 27. ARM 29. HIP 31. FDR 33. DDE
35. $-1, 2, -3$ 37. $-3, 19$

39. THE FIRST STEP TO WISDOM IS LEARNING THE PROPER NAMES OF THINGS (Confucius)

Chapter 5

Section 5.1

1. 3 3. 1, 2 5. 0, 3 7. 6 9. 0, 1, 2, 3, 4, 5, 6 11. $-9, -2, 5$
13. 5, 9 15. no solutions 17. no solutions 19. 28 21. 4, 16
23. 16, 100, 172, 256 25. $3y + 3 \equiv 0 \pmod 3$; $x = 1, 4, 7$
27. $27y + 7 \equiv 0 \pmod 5$; $x = 23$ 29. no solutions 31. 15, 105
33. $9x + 3$ 35. x

Section 5.2

1. $10x + 3$ 3. $70x^9 - 14x^6 + 3$ 5. $3x^2 + 6x + 3$ 7. 2, 5, 8
9. 5, 14, 23 11. no solutions 13. 5 15. 4 17. no solutions
21. 4, 29, 72, 97 23. 4, 372, 629, 997 25. 8, 14, 29, 35, 50, 56
27. 50, 59 37. $x = 97$ 39. $z = 21$ 45. 0, 1, 2, 3, 4, 5, 6, 7, 9

Section 5.3

1. 1, 2, 4 3. 1, 2, 4, 8, 9, 13, 15, 16 5. -1 7. -1 9. 1 11. 1
13. 1 15. 1 17. 1 19. 1, 5 21. no solutions 23. 12, 13
25. 2 27. 2 29. no solutions 31. 11, 16 33. no solutions
35. no solutions 41. $m = 8$, $a = 3$, $b = 5$ 43. T

Section 5.4

1. -1 3. 1 5. -1 7. 1 9. 1 11. undefined 13. 1
15. -1 17. 1 19. -1 21. -1 23. no 25. no 27. no
29. no 31. 0 33. 2 39.

h	1	2	3	4	5	6	7	8	9
x^*	5	-9	-4	1	6	-8	-3	2	7

41. 0, 0, 1, 1, 1, 1, 2, 2

Section 5.5

1. Al 3. Betty 5. 423 7. 162 9. 11, 12 11. 17, 26
13. 64, 202, 235, 373 15. 146, 284, 705, 843 17. 13, 20, 57, 64
19. 113, 135, 206, 228 23. 1, 1, 1, and 5, 5, 5 25. $p \equiv 3 \pmod 8$

Chapter 6

Section 6.1

1. 64, .716 3. -5, .667 5. 12, 0 7. 3, .99 9. yes
11. no; 2 doesn't match with a positive integer 13. yes 15. yes

17. $Y = .774\ldots$ 19. 3, 2/3 23. t matches with $(t + 2)/5$
25. t matches with $1 + \log_2 t$ 29. T 31. $X = 0$ 33. T 35. T

Section 6.2

1. −6, 7, −20; 1, −1, 3 3. 2, −9, 65; 1, −5, 36 5. $x = -32, y = 23$
7. $x = -4, y = 65$ 9. $x = 3, y = 1$ 11. 2, 7, 30; 1, 3, 13
13. −4, −3, −22; 1, 1, 7 15. 64/29 17. 79/248 19. $[1,1,3,1,1,4]$
21. $[-4,1,3,1,2]$ 23. $[3,7,7]$ 25. 4, 5, 14/3, 19/4, 71/15

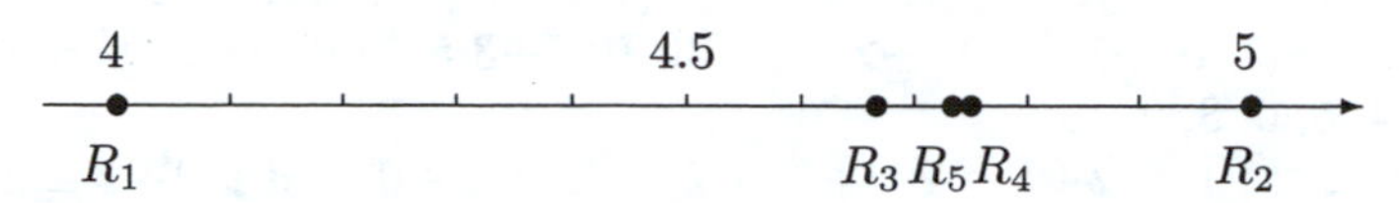

27. 3, 22/7, 157/50

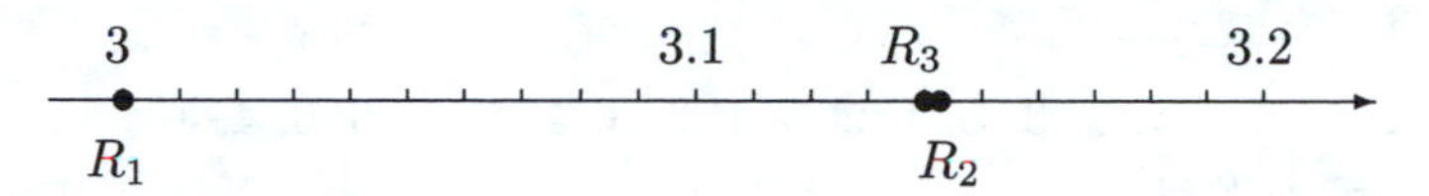

29. $h_4 = 1 + q_1q_2 + q_1q_4 + q_3q_4 + q_1q_2q_3q_4,\ k_4 = q_2 + q_4 + q_2q_3q_4$

Section 6.3

1. $(1+\sqrt{5})/2$ 3. $1+\sqrt{3}$ 5. $(\sqrt{13}-1)/2$ 7. $(\sqrt{15}-6)/3$ 9. $[0,\overline{1}]$
11. $[0,4,\overline{3}]$ 13. $[3,\overline{2,6}]$ 15. $[\overline{2,1,3}]$ 17. $[0,1,3,1,\ldots]$ 19. $[0,5,1,3,\ldots]$
27. within 1/742

Section 6.4

1. $.\overline{24}$ 3. $.3\overline{571428}$ 5. $.\overline{461538}$ 7. $1.55\overline{769230}$ 9. 7/9
11. 11/15 13. 61/45 15. 44/101 17. 6 19. 1 21. 4 23. no
25. yes 27. yes 29. 3 31. 3 33. none 35. 11, 33, 99

Section 6.5

1. 8 3. 2 5. 40 7. 0 9. 50, 53, 54, 58, 59 11. 4 13. 11

15.

a	1	2	3	4	5	6
index of a	0	2	1	4	5	3

17. $m = 3,\ a = b = 2$
19. $m = 2,\ a = b = 1$ 21. $m = 2,\ a = 1$ 23. T 25. $m = 7,\ c = 6$
29. A complete solution to $x^2 - 1 \equiv 0 \pmod 8$ has 4 elements.

Chapter 7

Section 7.1

1. $(24, 7, 25)$ 3. $(60, 11, 61)$ 5. $(12, 35, 37)$ 7. $(20, 21, 29)$
9. $u = 4,\ v = 3$ 11. $u = 9,\ v = 2$ 13. $u = 7,\ v = 6,\ z = 85$

15. $v = 8$, $x = 144$, $z = 145$ 17. $v = 5$, $x = 80$, $y = 39$
19. $u = 9$, $y = 77$, $z = 85$ 21. $(24, 7, 25)$, $(24, 143, 145)$
23. $(20, 21, 29)$, $(220, 21, 221)$ 25. $(44, 117, 125)$, $(117, 44, 125)$
27. $(16, 63, 65)$, $(16, 30, 34)$, $(16, 12, 20)$
29. $(12, 35, 37)$, $(612, 35, 613)$, $(120, 35, 125)$, $(84, 35, 91)$
33. $u = (n+1)/2$, $v = (n-1)/2$

Section 7.2

1. yes 3. no 5. yes 7. no 9. $(1\cdot1+3\cdot4)^2+(1\cdot4-3\cdot1)^2 = 13^2+1^2$
11. $(1\cdot1+2\cdot4)^2+(1\cdot4-2\cdot1)^2 = 9^2+2^2$ 13. $(0\cdot2+3\cdot5)^2+(0\cdot5-3\cdot2)^2 = 15^2+6^2$
15. 1, 2, 3, 4, 5, 6, 8, 9, 10
17. $0\cdot2+0 \equiv 2\cdot2+1$, $0\cdot2+1 \equiv 2\cdot2+2$, $0\cdot2+2 \equiv 1\cdot2+0$, and $1\cdot2+2 \equiv 2\cdot2+0$, all modulo 5.

Section 7.3

1. 0, 0, 2, 4; 1, 1, 3, 3 3. 0, 1, 2, 5; 1, 2, 3, 4 5. 0, 0, 4, 4
7. $x = 3$, $y = 1$, $k = 1$ 9. $x = 8$, $y = 2$, $k = 3$ 11. $8^2 + 2^2 + 0^2 + 3^2$
13. $8^2 + (-2)^2 + 1^2 + 1^2$ 15. $3p = 4^2 + 4^2 + 3^2 + 4^2$
17. $1p = (-3)^2 + (-1)^2 + (-1)^2 + 0^2$ 21. 24, 29, 32

Section 7.4

1. $8+8+1+1+1+1$ 3. $8+8+8+8+8$ 5. $64+27+8+1$
7. $81+16+1+1+1$ 9. $81+81+16+16+1+1+1+1+1+1$ 11. 23

Section 7.5

1. 17, 12; 99, 70 3. 49, 20; 485, 198 5. 199, 60; 3970, 1197
7. −3, 2, −3, 1 9. −3, 1, −3, 1 11. 2, 1; 14, 5 13. 9, 4; 161, 72
15. $(x, y, u, v, U, V) = (5, 1, 2, 1, 34, 7)$ 17. $(x, y, u, v, U, V) = (9, 4, 1, 2, 49, 22)$
19. 1, 3; 4, 2; 5, 1 21. 7, 1; 9, 3; 23, 11 23. 11, 3

Index